I0503913

# MIT INTEGRATION BEE

## Solutions Of Qualifying Tests From 2010 To 2023

### First Edition

Mohammad S. Alkousa

Copyright © 2023 by Mohammad S. Alkousa

All rights reserved. No part of this book may be reproduced, distributed, or trans-
mitted in any form or by any means, including photocopying, recording, or other
electronic or mechanical methods, without the prior written permission of the publisher, except as provided by United States of America copyright law. For permission
requests, e-mail the publisher at mohammad.alkousa@phystech.edu

The first edition published May 2023

*To my parents, my wife, and my lovely son Ali.*

# Preface

The Integration Bee is an annual integral calculus competition pioneered in 1981 at the Massachusetts Institute of Technology (MIT). Nowadays, Integration Bee contests are regularly conducted in major American universities, including the University of Florida, Florida Polytechnic University, the University of Scranton, Connecticut College, Central Connecticut State University, Columbia University, the State University of New York, the University of Wisconsin–Madison, Prairie State College (Illinois), the University of Illinois at Urbana–Champaign, University of Dayton (Ohio), Louisiana Tech University, the University of North Texas, Brigham Young University, Utah Valley University, Fresno State University, Cosumnes River College, the University of California, Berkeley, various other institutions in California, and Oregon State University. The Integration Bee is also held at the University of Cambridge, University of New South Wales, Australia, and the Indian Institute of Science Education and Research, Pune. The Bee held at Cambridge has participants from other British universities, including the University of Oxford, Imperial College London, Durham University, and the University of Warwick.

While integral calculus is no longer an actively researched topic in mathematics, there is some correlation between success in the integration bee and success in other areas of mathematics, where the Integration Bees share a common goal of challenging students to solve complex integration problems quickly and accurately. They provide opportunities for students to test their mathematical skills and compete with their peers. While the format and rules may vary, the Integration Bee competitions in the United States all share a commitment to promoting mathematical excellence and fostering a love of mathematics in students.

Despite the great popularity of MIT Integration Bee, there is no book that contains sufficient guidance and comprehensive and detailed solutions to the proposed integrals (or questions) each year. This created an incentive to solve these integrals (from 2010 to 2023). The solutions are in different ways, with some useful comments and recommendations. It is worth noting that there are individual efforts by some of those who offer solutions to these integrals on YouTube[1]. Also, there are sparse solutions for some different integrals in https://www.quora.com/.

---

[1]See for example the following channels: blackpenredpen, Owls Math, Math Solving Channel, Polar Pi, K.O. MATH., MathTuts

The main reason for presenting solutions to problems from 2010 to 2023 only, although the competition started in 1981, is the lack of access to issues that were addressed before 2010. On the official website of the MIT Integration Bee[2], there are only integrals that were given from 2010 to 2023, with the final answers for each question without any detail to the solution, although there are many challenging problems that require time and advanced calculus skills to solve. Also, while teaching students of all scientific disciplines the course of calculus, the search for various problems and integrals with multiple levels of challenges formed a great incentive to collect and solve these problems. The book will also be a good reference for teachers who teach calculus, as it provides them with many problems of different levels of difficulty, which are fun for students while teaching calculus. Therefore, this book is for students preparing for the Integration Bee, teachers looking for integration problems with different levels of challenge to use in their classroom, or math enthusiasts looking to expand their knowledge and skills in integration calculus, where this book has something for you.

This book consists of 15 chapters. In the first chapter, the most important ideas and techniques of integration were presented. We mentioned the basic integration formulas, integration by substitution, integration by parts, trigonometric and hyperbolic integrals, integrals of irrational functions, integration of binomial differentials, the definition of beta and gamma functions, and some additions such as the King property of integration, the Gaussian integral, Leibniz integral rule and Lebesgue's dominated convergence theorem. In the second chapter, the integrals that were given in the competition MIT Integration Bee from 2010 to 2023 were presented. In the remaining chapters, detailed solutions to the integrals of each year were presented in their own chapter.

I want to mention that this is the first edition of the book, which will be updated periodically every year, as solutions to problems will be added for the coming years. Finally, I would like to thank all of those who supported me in any respect during my work on this book. Many thanks to Ali Shadhar who provide me with suitable Latex files for publishing the book on Amazon, without his help it was very hard to publish this book with good quality. I would be grateful to everyone who will find any typos, addition, comment, or other remarkable methods to the solution and write me to mohammad.alkousa@phystech.edu. I would also be very grateful to everyone who can send questions that were presented in the competition before 2010 so that they can be solved in detail and included in another separate book.

Mohammad S. Alkousa
May 2023

---

[2]https://math.mit.edu/~yyao1/integrationbee.html

# Contents

# Chapter 1

# Review of the Fundamentals and Some Techniques of Integration

In this chapter, we will mention the fundamental rules and some techniques of integration, which are widely used in integration arithmetic competitions. For more details, there are many books and references on calculus, you can see for example [1, 2, 3].

## 1.1 Indefinite integrals, basic integration formulas

**Definition 1** *Let $F$ be a differentiable function on a given open interval $I$, and $f$ be a function defined on $I$. The function $F$ is called an antiderivative of a function $f$ on a given open interval if $F'(x) = f(x)$ for all $x \in I$.*

**Theorem 1** *If $F(x)$ is any antiderivative of $f(x)$ on an open interval $I$, then for any constant $C$ the function $F(x) + C$ is also an antiderivative on $I$. Moreover, each antiderivative of $f(x)$ on $I$ can be expressed in the form $F(x) + C$ by choosing the constant $C$ appropriately.*

The process of finding antiderivatives is called *antidifferentiation* or *integration*. Thus, if

$$\frac{d}{dx}F(x) = f(x),$$

then integrating (or antidifferentiating) the function $f(x)$ produces an antiderivative of the form $F(x) + C$, and we write

$$\int f(x)dx = F(x) + C,$$

1

where $C \in \mathbb{R}$ is an arbitrary constant.

## 1.1.1   Basic integration formulas

Here and in what follows, $C \in \mathbb{R}$ denotes any arbitrary constant.

**(1)**
$$\int a\,dx = ax + C, \quad a \in \mathbb{R}.$$

**(2)**
$$\int x^a\,dx = \frac{x^{a+1}}{a+1} + C, \quad a \in \mathbb{R} \setminus \{-1\}.$$

More generally, we have

$$\int (f(x))^a\, f'(x)dx = \frac{(f(x))^{a+1}}{a+1} + C, \quad a \in \mathbb{R} \setminus \{-1\}.$$

**(3)**
$$\int \frac{dx}{x} = \ln|x| + C, \quad x \neq 0.$$

More generally, we have

$$\int \frac{f'(x)}{f(x)}dx = \ln|f(x)| + C, \quad f(x) \neq 0.$$

**(4)**
$$\int \sin(ax + b)dx = -\frac{1}{a}\cos(ax + b) + C, \quad a \in \mathbb{R}^*, b \in \mathbb{R}.$$

**(5)**
$$\int \cos(ax + b)dx = \frac{1}{a}\sin(ax + b) + C, \quad a \in \mathbb{R}^*, b \in \mathbb{R}.$$

**(6)**
$$\int \frac{dx}{\cos^2(ax + b)} = \int \sec^2(ax + b)dx$$
$$= \frac{1}{a}\tan(ax + b) + C, \quad a \in \mathbb{R}^*, b \in \mathbb{R}.$$

**(7)**
$$\int \frac{dx}{\sin^2(ax + b)} = \int \csc^2(ax + b)dx$$
$$= -\frac{1}{a}\cot(ax + b) + C, \quad a \in \mathbb{R}^*, b \in \mathbb{R}.$$

**(8)**

$$\int \sinh(ax+b)dx = \frac{1}{a}\cosh(ax+b) + C, \quad a \in \mathbb{R}^*, b \in \mathbb{R}.$$

**(9)**

$$\int \cosh(ax+b)dx = \frac{1}{a}\sinh(ax+b) + C, \quad a \in \mathbb{R}^*, b \in \mathbb{R}.$$

**(10)**

$$\int \frac{dx}{\cosh^2(ax+b)} = \frac{1}{a}\tanh(ax+b) + C, \quad a \in \mathbb{R}^*, b \in \mathbb{R}.$$

**(11)**

$$\int \frac{dx}{\sinh^2(ax+b)} = -\frac{1}{a}\coth(ax+b) + C, \quad a \in \mathbb{R}^*, b \in \mathbb{R}.$$

**(12)**

$$\int a^{\alpha x+\beta}dx = \frac{a^{\alpha x+\beta}}{\alpha \ln(a)} + C, \quad a \in ]0,\infty[\setminus\{1\}, \alpha \in \mathbb{R}^*, \beta \in \mathbb{R}.$$

Special case: when $a = e$, then we get

$$\int e^{\alpha x+\beta}dx = \frac{1}{\alpha}e^{\alpha x+\beta} + C.$$

**(13)**

$$\int \frac{dx}{a^2+x^2} = \frac{1}{a}\arctan\left(\frac{x}{a}\right) + C, \quad a \in \mathbb{R}^*.$$

**(14)**

$$\int \frac{dx}{a^2-x^2} = \frac{1}{a}\operatorname{artanh}\left(\frac{x}{a}\right) + C = \frac{1}{2a}\ln\left|\frac{a+x}{a-x}\right| + C, \quad a \in \mathbb{R}^*.$$

**(15)**

$$\int \frac{dx}{\sqrt{a^2-x^2}} = \arcsin\left(\frac{x}{a}\right) + C = -\arccos\left(\frac{x}{a}\right) + C, \quad x < a.$$

**(16)**

$$\int \frac{dx}{\sqrt{x^2+a^2}} = \ln\left|x+\sqrt{x^2+a^2}\right| + C = \operatorname{arsinh}\left(\frac{x}{a}\right) + C, \quad a > 0.$$

**(17)**

$$\int \frac{dx}{\sqrt{x^2-a^2}} = \ln\left|x+\sqrt{x^2-a^2}\right| + C = \operatorname{arcosh}\left(\frac{x}{a}\right) + C, \quad x > a > 0.$$

**(18)**

$$\int \frac{dx}{\cos x} = \int \sec x\, dx = \ln|\sec x + \tan x| + C = \ln\left|\tan\left(\frac{\pi}{4} + \frac{x}{2}\right)\right| + C.$$

**(19)**

$$\int \frac{dx}{\sin x} = \int \csc x\, dx = -\ln|\csc x + \cot x| + C = \ln\left|\tan\left(\frac{x}{2}\right)\right| + C.$$

**(20)** Let $a, b \in \mathbb{R}$, such that $a^2 + b^2 \neq 0$. Then

$$I := \int e^{ax}\cos(bx)dx = \frac{e^{ax}}{a^2 + b^2}\left(a\cos(bx) + b\sin(bx)\right) + C, \qquad (1.1)$$

where $C \in \mathbb{R}$.

Indeed, by using the integration by parts (see Section 1.4), let us assume

$$u = e^{ax} \implies du = ae^{ax}dx, \quad dv = \cos(bx)dx \implies v = \frac{1}{b}\sin(bx).$$

Thus, we have

$$I = \frac{1}{b}e^{ax}\sin(bx) - \frac{a}{b}\underbrace{\int e^{ax}\sin(bx)dx}_{:=J}.$$

For the integral $J$, by using the integration by parts, let us assume

$$u = e^{ax} \implies du = ae^{ax}dx, \quad dv = \sin(bx)dx \implies v = -\frac{1}{b}\cos(bx).$$

Thus, we have

$$I = \frac{1}{b}e^{ax}\sin(bx) - \frac{a}{b}\left(-\frac{1}{b}e^{ax}\cos(bx) + \frac{a}{b}\underbrace{\int e^{ax}\cos(bx)dx}_{I}\right)$$

$$= \frac{1}{b}e^{ax}\sin(bx) + \frac{a}{b^2}e^{ax}\cos(bx) - \frac{a^2}{b^2}I.$$

Thus, we get

$$\left(1 + \frac{a^2}{b^2}\right)I = e^{ax}\left(\frac{b\sin(bx) + a\cos(bx)}{b^2}\right).$$

Therefore, we have

$$I = \frac{e^{ax}}{a^2 + b^2} \left( a \cos(bx) + b \sin(bx) \right) + C.$$

**(21)** In similar of the previous, we get

$$\int e^{ax} \sin(bx) dx = \frac{e^{ax}}{a^2 + b^2} \left( a \sin(bx) - b \cos(bx) \right) + C, \qquad (1.2)$$

where $C \in \mathbb{R}$.

### 1.1.2  Properties of indefinite integrals

(1)
$$\left( \int f(x) dx \right)' = f(x).$$

(2)
$$d \left( \int f(x) dx \right) = f(x) dx.$$

(3)
$$\int dF(x) = F(x) + C, \quad C \in \mathbb{R}.$$

(4) Let $f_1, \ldots, f_n$ be integrable functions, then

$$\int (f_1(x) + \ldots + f_n(x)) \, dx = \int f_1(x) dx + \ldots + \int f_n(x) dx.$$

(5)
$$\int a f(x) dx = a \int f(x) dx, \quad \forall a \in \mathbb{R}^*.$$

## 1.2  Definite integrals

**Theorem 2 (The Fundamental Theorem of Calculus, Part 1)** *If $f$ is continuous on $[a, b]$ and $F$ is any antiderivative of $f$ on $[a, b]$ (i.e. $F'(x) = f(x)$, $\forall x \in [a, b]$), then*

$$\int_a^b f(x) dx = \int_a^b F'(x) dx = F(b) - F(a) := [F(x)]_a^b.$$

**Theorem 3** *If a function f is continuous on an interval $[a, b]$, then f is integrable on $[a, b]$, and the area A between the graph of f and the interval $[a, b]$ is*

$$A = \int_a^b f(x)dx.$$

## 1.2.1  Properties of definite integrals

**(1)** If $a$ is in the domain of $f$, then

$$\int_a^a f(x)dx = 0.$$

**(2)** If $f$ is integrable on $[a, b]$, then

$$\int_a^b f(x)dx = - \int_b^a f(x)dx.$$

**(3)** If $f$ is integrable on $[a, b]$ and if $\alpha \in \mathbb{R}$ is a constant, then

$$\int_a^b \alpha f(x)dx = \alpha \int_a^b f(x)dx.$$

**(4)** If $f_1, \ldots, f_n$ are integrable functions on $[a, b]$, then

$$\int_a^b (f_1(x) + \ldots + f_n(x)) \, dx = \int_a^b f_1(x)dx + \ldots + \int_a^b f_n(x)dx.$$

**(5)** If $f$ is integrable on a closed interval containing the three points $a, b, c$ such that $a \leq c \leq b$, then

$$\int_a^b f(x)dx = \int_a^c f(x)dx + \int_c^b f(x)dx.$$

**(6)** The variable of integration in a definite integral plays no role in the end result, it is often referred to as a *dummy variable*. This means

$$\int_a^b f(x)dx = \int_a^b f(t)dt.$$

## 1.3   Integration by substitution

The substitution formula is often written in the form

$$\int f(u)du = \int f(g(x))g'(x)dx, \quad \text{where } u = g(x). \tag{1.3}$$

For definite integrals, the formula (1.3), will has the following form

$$\int_a^b f(g(x))g'(x)dx = \int_{g(a)}^{g(x)} f(u)du. \tag{1.4}$$

## 1.4   Integration by parts

Be attentive!

$$\int f(x)\,g(x)dx \neq \int f(x)dx \int g(x)dx.$$

But, if $f'$ and $g'$ are continuous functions[1] of $x$, then the integration by parts formula has the following form

$$\int f(x)g'(x)dx = f(x)g(x) - \int f'(x)g(x)dx. \tag{1.5}$$

Sometimes it is easier to remember the formula (1.5) if we write it in differential form. Let $u = f(x)$ and $v = g(x)$. Then $du = f'(x)dx$ and $dv = g'(x)dx$. Using the substitution rule, the integration by parts formula (1.5) becomes

$$\int udv = uv - \int vdu. \tag{1.6}$$

This formula expresses one integral $\int u\,dv$, in terms of a second integral $\int vdu$. With a proper choice of $u$ and $v$, the second integral may be easier to evaluate than the first. In using the formula (1.6), various choices may be available for $u$ and $dv$. For the definite integrals, we have

$$\int_a^b udv = [uv]_a^b - \int_a^b vdu.$$

From (1.6), we can see that the Integration by parts is a technique for simplifying integrals of the multiplication of two functions. It is useful when one of these functions can be differentiated repeatedly and the second can be integrated repeatedly without difficulty.

---

[1]See: Alexander Kheifets and James Propp: A Counterexample to Integration by Parts. Mathematics Magazine, Vol. 83, No. 3 (June 2010), pp. 222-225. Where in this paper, the authors exhibit two differentiable functions $f$ and $g$ for which the function $f'g$ and $fg'$ are not integrable, so that the integration by parts formula does not apply.

Let us mention some of the integrals that can be calculated by using the integration by parts.

**(1)** The integrals

$$\int P_n(x)\sin(ax+b)dx, \quad \int P_n(x)\cos(ax+b)dx, \quad \int P_n(x)e^{ax+b}dx,$$

$$\int P_n(x)\sinh(ax+b)dx, \quad \int P_n(x)\cosh(ax+b)dx,$$

where $P_n(x)$ is a polynomial of degree $n$ and $a, b \in \mathbb{R}$.

For these integrals, we assume $u = P_n(x)$ and the remainder is $dv$. By using the integration by parts $n$ times we calculate the given integral.

**(2)** The integrals

$$\int P_n(x)\ln^m xdx, \quad \int \arctan xdx, \quad \int \text{arccot}\, xdx, \quad \int \arcsin xdx,$$

$$\int \arccos xdx, \quad \int \text{arsinh}\, xdx, \quad \int \text{arcosh}\, xdx.$$

For the first integral, we assume $u = \ln^m x$ and by using the integration by parts $m$ times we calculate this integral. For other integrals, we assume $dv = dx$ and the remainder (i.e. the inverse trigonometric or inverse hyperbolic functions) is $u$.

## 1.5   Trigonometric and hyperbolic integrals

### 1.5.1   Basic identities for trigonometric and inverse trigonometric functions

Remember that, for any $x \in \mathbb{R}$ and $k \in \mathbb{Z}$, we have

$$\sin(-x) = -\sin x, \quad \cos(-x) = \cos x,$$

$$\tan(-x) = -\tan x, \quad \cot(-x) = -\cot x.$$

$$\sin(x + 2\pi k) = \sin x, \quad \cos(x + 2\pi k) = \cos x,$$

$$\tan(x + \pi k) = \tan x, \quad \cot(x + \pi k) = \cot x.$$

## Double-angle formulae

$$\sin(2x) = 2\sin x \cos x = \frac{2\tan x}{1 + \tan^2 x},$$

$$\cos(2x) = \cos^2 x - \sin^2 x = 2\cos^2 x - 1 = 1 - 2\sin^2 x = \frac{1 - \tan^2 x}{1 + \tan^2 x},$$

$$\tan(2x) = \frac{2\tan x}{1 - \tan^2 x}.$$

## Sum and difference formulas

$$\sin(x \pm y) = \sin x \cos y \pm \cos x \sin y,$$

$$\cos(x \mp y) = \cos x \cos y \pm \sin x \sin y,$$

$$\tan(x \pm y) = \frac{\tan x \pm \tan y}{1 \mp \tan x \tan y}.$$

## Formulas for transforming the product into sum or difference

$$\cos(mx)\cos(nx) = \frac{1}{2}\left[\cos(m+n)x + \cos(m-n)x\right],$$

$$\sin(mx)\cos(nx) = \frac{1}{2}\left[\sin(m+n)x + \cos(m-n)x\right],$$

$$\sin(mx)\sin(nx) = -\frac{1}{2}\left[\cos(m+n)x - \cos(m-n)x\right].$$

## Inverse trigonometric functions

For the domains and ranges of the inverse trigonometric functions, we have

$$\arcsin : [-1, 1] \longrightarrow [-\pi/2, \pi/2], \quad \arccos : [-1, 1] \longrightarrow [0, \pi],$$
$$\arctan : \mathbb{R} \longrightarrow (-\pi/2, \pi/2), \quad \operatorname{arccot} : \mathbb{R} \longrightarrow (0, \pi).$$

For these functions, we have

$$\arcsin(-x) = -\arcsin x, \quad \arccos(-x) = \pi - \arccos x,$$

$$\arctan(-x) = -\arctan x, \quad \operatorname{arccot}(-x) = \pi - \operatorname{arccot} x.$$

## Relationships between trigonometric functions and inverse trigonometric functions

**(1)** If $\theta = \arcsin x$, where $x \in [-1, 1]$, then we can get the following figure

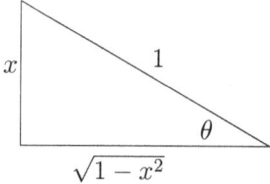

and

$$\sin(\arcsin x) = x, \quad \cos(\arcsin x) = \sqrt{1 - x^2}, \quad \tan(\arcsin x) = \frac{x}{\sqrt{1 - x^2}}.$$

(2) If $\theta = \arccos x$, where $x \in [-1, 1]$, then we can get the following figure

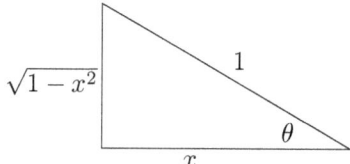

and

$$\sin(\arccos x) = \sqrt{1 - x^2}, \quad \cos(\arccos x) = x, \quad \tan(\arccos x) = \frac{\sqrt{1 - x^2}}{x}.$$

(3) If $\theta = \arctan x$, where $x \in \mathbb{R}$, then we can get the following figure

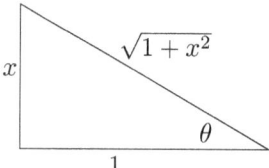

and

$$\sin(\arctan x) = \frac{x}{\sqrt{1 + x^2}}, \quad \cos(\arctan x) = \frac{1}{\sqrt{1 + x^2}}, \quad \tan(\arctan x) = x.$$

(4) If $\theta = \text{arccot } x$, then we can get the following figure

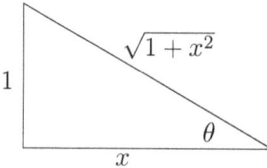

and

$$\sin(\text{arccot } x) = \frac{1}{\sqrt{1 + x^2}} \quad \forall x \in \mathbb{R}, \quad \cos(\text{arccot } x) = \frac{x}{\sqrt{1 + x^2}} \quad \forall x \in \mathbb{R},$$

$$\tan(\operatorname{arccot} x) = \frac{1}{x} \quad \forall x \in \mathbb{R} \setminus \{0\}.$$

### 1.5.2 Trigonometric integrals

Let $R$ be a rational function in its arguments (maybe more than one variable). For integrals

$$\int R\,(\sin x, \cos x)\,dx, \tag{1.7}$$

we can use the substitution $\tan\left(\frac{x}{2}\right) = t$, thus $x = 2\arctan t$, and $dx = \frac{2dt}{1+t^2}$. Also, we have

$$\sin x = \frac{2\sin\left(\frac{x}{2}\right)\cos\left(\frac{x}{2}\right)}{1} = \frac{2\sin\left(\frac{x}{2}\right)\cos\left(\frac{x}{2}\right)}{\sin^2\left(\frac{x}{2}\right) + \cos^2\left(\frac{x}{2}\right)} = \frac{2\tan\left(\frac{x}{2}\right)}{1 + \tan^2\left(\frac{x}{2}\right)} = \frac{2t}{1+t^2},$$

$$\cos x = \frac{\cos^2\left(\frac{x}{2}\right) - \sin^2\left(\frac{x}{2}\right)}{1} = \frac{\cos^2\left(\frac{x}{2}\right) - \sin^2\left(\frac{x}{2}\right)}{\cos^2\left(\frac{x}{2}\right) + \sin^2\left(\frac{x}{2}\right)} = \frac{1 - \tan^2\left(\frac{x}{2}\right)}{1 + \tan^2\left(\frac{x}{2}\right)} = \frac{1-t^2}{1+t^2}.$$

This gives the following integral

$$\int R\left(\frac{2t}{1+t^2}, \frac{1-t^2}{1+t^2}\right)\frac{2dt}{1+t^2},$$

which is an integral of a rational function in $t$.

For the trigonometric integrals (1.7), there are some special cases that can be listed as follows.

**(1)** For the integrals

$$\int R\,(\cos x)\sin x dx,$$

we can use the substitution $\cos x = t$, so that $-\sin x dx = dt$. This gives the integral $-\int R(t)dt$, which is an integral of the rational function in $t$.

**Example 1** *For the integral*

$$I = \int \frac{\sin^3 x}{2 + \cos x}dx,$$

*we have*

$$I = \int \frac{\sin^2 x \sin x}{2 + \cos x}dx = \int \frac{1 - \cos^2 x}{2 + \cos x}\sin x dx.$$

*Let us use the substitution $\cos x = t$, which gives the following*

$$I = -\int \frac{1-t^2}{1+t^2}dt = \int \left(t - 2 + \frac{3}{2+t}\right)dt = \frac{t^2}{2} - 2t + 3\ln|2 + t| + C$$

$$= \frac{1}{2} \cos^2 x - \cos x + 3 \ln |2 + \cos x| + C,$$

where $C \in \mathbb{R}$.

**(2)** For the integrals

$$\int R \left( \sin x \right) \cos x \, dx,$$

we can use the substitution $\sin x = t$, so that $\cos x \, dx = dt$. This gives the integral $\int R(t) dt$, which is an integral of the rational function in $t$.

**(3)** For the integrals

$$\int R \left( \tan x \right) dx,$$

we can use the substitution $\tan x = t$, so that $x = \arctan t$ and $dx = \frac{dt}{1+t^2}$. This gives the integral $\int R(t) \frac{dt}{1+t^2}$, which is an integral of the rational function in $t$.

**(4)** For the integrals

$$\int R \left( \sin^m x, \cos^n x \right) dx,$$

where the powers $m, n$ of $\sin x$ and $\cos x$ are even, we can use the substitution

$$\tan x = t \implies x = \arctan t \implies dx = \frac{dt}{1+t^2},$$

and

$$\cos^2 x = \frac{1}{1 + \tan^2 x} = \frac{1}{1+t^2}, \quad \sin^2 x = \frac{\tan^2 x}{1 + \tan^2 x} = \frac{t^2}{1+t^2}.$$

This gives an integral of a rational function in $t$.

**Example 2** *For the integral*

$$I = \int \frac{dx}{2 - \sin^2 x},$$

*with* $\tan x = t$*, we have*

$$I = \int \frac{dt}{\left(1 + t^2\right) \left(2 - \frac{t^2}{1+t^2}\right)} = \int \frac{dt}{2 + t^2} = \frac{1}{\sqrt{2}} \arctan \left( \frac{t}{\sqrt{2}} \right) + C$$

$$= \frac{1}{\sqrt{2}} \arctan \left( \frac{\tan x}{\sqrt{2}} \right) + C,$$

where $C \in \mathbb{R}$.

**(5)** For the integrals

$$\int \sin^m x \cos^n x dx \quad (m, n \in \mathbb{Z}), \tag{1.8}$$

we can list the following cases.

- One of the numbers $m, n$ is an odd positive (for example $n$ is this number). Then by setting $n = 2p + 1$, where $p \in \mathbb{Z}^+$, we get

$$I = \int \sin^m x \cos^{2p+1} x dx = \int \sin^m x \cos^{2p} x \cos x dx$$

$$= \int \sin^m x \left(1 - \sin^2 x\right)^p \cos x dx.$$

Thus, when we use the substitution $\sin x = t$, we get the following integral

$$I = \int t^m (1 - t^2)^p dt,$$

which is an integral of a rational function when $m < 0$. If $m > 0$, then the integrand will be a polynomial with degree $m + 2p$.

**Example 3** *For the integral*

$$I = \int \frac{\cos^3 x}{\sin^4 x} dx,$$

*we have*

$$I = \int \frac{\cos^2 x \cos x}{\sin^4 x} dx = \int \frac{(1 - \sin^2 x)}{\sin^4 x} \cos x dx.$$

*Let us use the substitution $\sin x = t$, thus $\cos x dx = dt$. This gives the following*

$$I = \int \frac{1 - t^2}{t^4} dt = \int \left(t^{-4} - t^{-2}\right) dt = -\frac{1}{3t^3} + \frac{1}{t} + C$$

$$= -\frac{1}{3 \sin^3 x} + \frac{1}{\sin x} + C,$$

*where $C \in \mathbb{R}$.*

- Both numbers $m$ and $n$ are even. Let $m = 2p$ and $n = 2q$. We have

$$\sin^2 x = \frac{1}{2} - \frac{1}{2} \cos(2x), \quad \cos^2 x = \frac{1}{2} + \frac{1}{2} \cos(2x). \tag{1.9}$$

Thus, we have the following integral

$$\int \sin^{2p} x \cos^{2q} x dx = \int \left(\frac{1}{2} - \frac{1}{2}\cos(2x)\right)^p \left(\frac{1}{2} + \frac{1}{2}\cos(2x)\right)^q dx.$$

After doing some simple calculations, we get even or odd powers for $\cos(2x)$. For the terms with odd powers, we return to the first case. For the terms with even powers, we can use (1.9) which gives an integral of the form $\int \cos(\alpha x)\, dx$, where $\alpha \in \mathbb{R}^*$.

**Example 4** *For the integral*

$$I = \int \sin^4 x dx,$$

*we have*

$$I = \int \sin^4 x dx = \frac{1}{4}\int (1 - \cos(2x))^2 dx$$

$$= \frac{1}{4}\int \left(1 - 2\cos(2x) + \cos^2(2x)\right) dx$$

$$= \frac{1}{4}\int \left(1 - 2\cos(2x) + \frac{1}{2} + \frac{1}{2}\cos(4x)\right) dx$$

$$= \frac{1}{4}\int \left(\frac{3}{2} - 2\cos(2x) + \frac{1}{2}\cos(4x)\right) dx$$

$$= \frac{3}{8}x - \frac{1}{4}\sin(2x) + \frac{1}{32}\cos(4x) + C,$$

*where $C \in \mathbb{R}$.*

- Both numbers $m, n$ are even and one of them is negative. In this case, we can use the substitution $\tan x = t$.

**Example 5** *For the integral*

$$I = \int \frac{\sin^2 x}{\cos^6 x} dx,$$

*we have*

$$I = \int \frac{\sin^2 x}{\cos^2 x} \frac{1}{\cos^2 x} \frac{dx}{\cos^2 x}.$$

*Let us use the substitution $\tan x = t$, thus $\frac{dx}{\cos^2 x} = dt$. This gives the following*

$$I = \int t^2(1 + t^2) dt = \int (t^4 + t^2) dt = \frac{t^5}{5} + \frac{t^3}{3} + C$$

$$= \frac{\tan^5 x}{5} + \frac{\tan^3 x}{3} + C,$$

*where $C \in \mathbb{R}$.*

## 1.5.3 Hyperbolic integrals

Hyperbolic integrals can be calculated in a similar way as trigonometric integrals. Here we mention some useful formulas for hyperbolic functions.

**Definitions**

$$\sinh x = \frac{e^x - e^{-x}}{2}, \quad \cosh x = \frac{e^x + e^{-x}}{2},$$

$$\tanh x = \frac{\sinh x}{\cosh x} = \frac{e^x - e^{-x}}{e^x + e^{-x}}, \quad \coth x = \frac{1}{\tanh x} = \frac{\sinh x}{\cosh x} = \frac{e^x + e^{-x}}{e^x - e^{-x}}.$$

**Fundamental formulas**

$$\cosh^2 x - \sinh^2 x = 1, \quad \sinh(-x) = -\sinh x,$$

$$\cosh(-x) = \cosh x, \quad \tanh(-x) = -\tanh x,$$

$$\sinh(a \pm b) = \sinh a \cosh b \pm \cosh a \sinh b,$$

$$\cosh(a \pm b) = \cosh a \cosh b \pm \sinh a \sinh b,$$

$$\tanh(a \pm b) = \frac{\tanh a \pm \tanh b}{1 \pm \tanh a \tanh b},$$

$$\cosh(2x) = \cosh^2 x + \sinh^2 x = 2\sinh^2 x + 1 = 2\cosh^2 x - 1,$$

$$\cosh^2 x = \frac{\cosh(2x) + 1}{2}, \quad \sinh^2 x = \frac{\cosh(2x) - 1}{2},$$

$$\sinh(2x) = 2\sinh x \cosh x, \quad \tanh(2x) = \frac{2\tanh x}{1 + \tanh^2 x}.$$

**Formulas for transforming the product into sum or difference**

$$\sinh(mx)\cosh(nx) = \frac{1}{2}\left[\sinh(m+n)x + \sinh(m-n)x\right],$$

$$\cosh(mx)\cosh(nx) = \frac{1}{2}\left[\cosh(m+n)x + \cosh(m-n)x\right],$$

$$\sinh(mx)\sinh(nx) = -\frac{1}{2}\left[\cosh(m+n)x - \cosh(m-n)x\right].$$

**Inverse hyperbolic functions**

$$\operatorname{arsinh} x = \ln\left(x + \sqrt{x^2 + 1}\right); \quad \forall x \in \mathbb{R},$$

$$\operatorname{arcosh} x = \ln\left(x + \sqrt{x^2 - 1}\right); \quad \forall x \in [1, \infty[,$$

$$\operatorname{artanh} x = \frac{1}{2} \ln \left( \frac{1+x}{1-x} \right) ; \qquad \forall x \in\, ]-1, 1[,$$

$$\operatorname{arcoth} x = \frac{1}{2} \ln \left( \frac{1+x}{1-x} \right) ; \qquad \forall x \in\, ]-\infty, -1[\, \cup\, ]1, \infty[.$$

Also, we have the following

$$\sinh(\operatorname{arcosh} x) = \sqrt{x^2 - 1}; \quad |x| > 1,$$

$$\sinh(\operatorname{artanh} x) = \frac{x}{\sqrt{1 - x^2}}; \quad -1 < x < 1,$$

$$\cosh(\operatorname{arsinh} x) = \sqrt{1 + x^2}; \quad x \in \mathbb{R},$$

$$\cosh(\operatorname{artanh} x) = \frac{1}{\sqrt{1 - x^2}}; \quad -1 < x < 1,$$

$$\tanh(\operatorname{arsinh} x) = \frac{x}{\sqrt{1 + x^2}}; \quad x \in \mathbb{R},$$

$$\tanh(\operatorname{arcosh} x) = \frac{\sqrt{x^2 - 1}}{x}; \quad |x| > 1.$$

## 1.5.4   Trigonometric and hyperbolic substitutions

Remember that

$$\sin^2 x + \cos^2 x = 1, \quad \cosh^2 x - \sinh^2 x = 1, \quad 1 + \tan^2 x = \frac{1}{\cos^2 x} = \sec^2 x.$$

Therefore

- To calculate

$$\int R \left( x, \sqrt{a^2 - b^2 x^2} \right) dx,$$

  where $a, b \in \mathbb{R}^*$ and $R(\cdot, \cdot)$ is a rational function in its arguments, we can use the substitution $x = \frac{a}{b} \sin t$, or $x = \frac{a}{b} \cos t$, or $x = \frac{a}{b} \tanh t$.

- To calculate

$$\int R \left( x, \sqrt{a^2 + b^2 x^2} \right) dx,$$

  where $a, b \in \mathbb{R}^*$, and $R(\cdot, \cdot)$ is a rational function in its arguments, we can use the substitution $x = \frac{a}{b} \tan t$, or $x = \frac{a}{b} \sinh t$.

- To calculate

$$\int R \left( x, \sqrt{b^2 x^2 - a^2} \right) dx,$$

  where $a, b \in \mathbb{R}^*$, and $R(\cdot, \cdot)$ is a rational function in its arguments, we can use the substitution $x = \frac{a}{b} \sec t$, or $x = \frac{a}{b} \cosh t$.

## 1.6  Integrals of irrational functions

For the integrals

$$\int R\left(x,\left(\frac{ax+b}{cx+d}\right)^{\frac{m}{n}},\ldots,\left(\frac{ax+b}{cx+d}\right)^{\frac{r}{s}}\right)dx,$$

where $m,n,\ldots,r,s\in\mathbb{Z}^*$, $R$ is a rational function in its arguments, and $a,b,c,d\in\mathbb{R}$ such that $ad-bc\neq0$. We can use the substitution $\frac{ax+b}{cx+d}=t^k$, where $k$ is a common denominator of the denominators $n,\ldots,s$. This substitution yields an integral of a rational function.

**Special case 1:** The integrals that have the following form

$$\int R\left(x,x^{\frac{m}{n}},\ldots,x^{\frac{r}{s}}\right)dx.$$

**Special case 1:** The integrals that have the following form

$$\int R\left(x,(ax+b)^{\frac{m}{n}},\ldots,(ax+b)^{\frac{r}{s}}\right)dx.$$

For the integrals

$$I=\int\frac{dx}{(x-\alpha)^k\sqrt{ax^2+bx+c}}. \tag{1.10}$$

we use the substitution $\frac{1}{x-\alpha}=t$, so that

$$dx=-\frac{dt}{t^2},\quad ax^2+bx+c=\frac{\left(a\alpha^2+b\alpha+c\right)t^2+(2a\alpha+b)t+a}{t^2}.$$

This gives the following integral

$$I=-\int\frac{t^{k-1}}{\sqrt{\left(a\alpha^2+b\alpha+c\right)t^2+(2a\alpha+b)t+a}}dt,$$

which can be calculated simpler than the given integral (1.10).

## 1.7  Integration of binomial differentials

An expression of the form $x^m\left(a+bx^n\right)^p$, where $a,b\in\mathbb{R}^*$ and $m,n,p\in\mathbb{Q}$ is called a binomial differential.

**Theorem 4** *The integral of a binomial differential*

$$I=\int x^m\left(a+bx^n\right)^p\,dx,$$

*is reduced to an integral of a rational function and thus is expressed in terms of elementary functions in the following three cases*

1. *When $p \in \mathbb{Z}$. In this case, to calculate $I$, we use the substitution $x = t^k$, where $k$ is a common denominator of the fractions $m, n$.*

2. *When $\frac{m+1}{n} \in \mathbb{Z}$. In this case, to calculate $I$, we use the substitution $a + bx^n = t^r$, where $r$ is the denominator of $p$.*

3. *When $\frac{m+1}{n} + p \in \mathbb{Z}$. In this case, we can write*

$$I = \int x^m \, x^{np} \left( b + ax^{-n} \right)^p \, dx = \int x^{m+np} \left( b + ax^{-n} \right)^p \, dx. \qquad (1.11)$$

*Let $m_1 := m + np$, $n_1 := -n$, $p_1 := p$. We note that*

$$\frac{m_1 + 1}{n_1} = \frac{m + np + 1}{-n} = -\left( \frac{m+1}{n} + p \right) \in \mathbb{Z}.$$

*Thus, the integral (1.11), is an integral of a binomial differential from the second case.*

## 1.8   Beta and Gamma functions

The gamma and beta functions are two important mathematical functions that arise in various areas of mathematics, physics, and engineering. In this section, we mention the definitions of these functions.

### 1.8.1   Gamma function

This function is also called *Euler's gamma function* and is defined as follows

$$\Gamma(a) = \int_0^\infty x^{a-1} e^{-x} dx, \quad \forall a > 0. \qquad (1.12)$$

For this function we have $\Gamma(a+1) = a\Gamma(a)$, and from this relation we get $\Gamma(n+1) = n!$, $\forall n \in \mathbb{N}$. Therefore, we see that the gamma function extends the factorial function from the non-negative integers to other real values.

From (1.12) and Gaussian integral (see (1.14)), we find the following important number $\Gamma(1/2) = \sqrt{\pi}$.

### 1.8.2 Beta function

For any $a > 0$ and $b > 0$, the beta function (also called the *Euler integral of the first kind*) is defined as follows

$$\beta(a, b) = \int_0^1 x^{a-1}(1 - x)^{b-1}dx.$$

The close relationship to the gamma function is

$$\beta(a, b) = \frac{\Gamma(a)\,\Gamma(b)}{\Gamma(a + b)}.$$

The trigonometric form of the beta function is

$$\beta(a, b) = 2 \int_0^{\pi/2} (\sin x)^{2a-1}(\cos x)^{2b-1}dx.$$

## 1.9 Additionals

### 1.9.1 King property of integration

This property states

$$\int_a^b f(x)dx = \int_a^b f(a + b - x)dx. \tag{1.13}$$

### 1.9.2 The Gaussian integral

The Gaussian integral (or *Euler–Poisson integral*) has the following form

$$\int_{-\infty}^{\infty} e^{-x^2}\,dx = \sqrt{\pi}. \tag{1.14}$$

### 1.9.3 Leibniz integral rule

Leibniz's integral rule provides us a way to differentiate an integral without first calculating the integral in the first place. This rule states the following:

Let $f(x, t)$ be a function such that both $f(x, t)$ and its partial derivative $f_x(x, t) = \frac{\partial}{\partial x} f(x, t)$ are continuous in $t$ and $x$ in some region of the $xt$-plane, including $a(x) \leq t \leq b(x)$, $x_0 \leq x \leq x_1$, where $x_0, x_1 \in \mathbb{R}$. Also, suppose that the functions $a(x)$ and $b(x)$ are both continuous and both have continuous derivatives for $x_0 \leq x \leq x_1$. Then, for $x_0 \leq x \leq x_1$,

$$\frac{d}{dx}\left(\int_{a(x)}^{b(x)} f(x, t)dt\right) = f(x, b(x))b'(x) - f(x, a(x))a'(x) + \int_{a(x)}^{b(x)} f_x(x, t)dt.$$

In the special case where the functions $a(x) = a$ and $b(x) = b$ are constants with values that do not depend on $x$, this simplifies to:

$$\frac{d}{dx}\left(\int_a^b f(x,t)dt\right) = \int_a^b f_x(x,t)dt. \tag{1.15}$$

### 1.9.4   Lebesgue's dominated convergence theorem

This theorem states that [4, 5]: if $f_n : \mathbb{R} \longrightarrow \mathbb{R}$ is a sequence of measurable functions such that the pointwise limit $\lim_{n\to\infty} f_n(x) := f(x)$ exists, and if there exists an integrable function $g$ such that $|f_n(x)| \leqslant g(x)$ for all $n$ and for all $x$, then $f$ is integrable and

$$\int_\mathbb{R} f(x)dx = \lim_{n\to\infty} \int_\mathbb{R} f_n(x)dx.$$

# Chapter 2

# The Questions of MIT Integration Bee, Qualifying Tests

## 2.1 2010 MIT Integration Bee, Qualifying Test

1. $\quad I_1 = \displaystyle\int_0^{\pi/2} \sin x \sin(2x) \sin(3x) dx.$

2. $\quad I_2 = \displaystyle\int_0^{\pi/2} \sin^3(2x) \cos x dx.$

3. $\quad I_3 = \displaystyle\int (x+1)^2 (x-1)^{1/3} dx.$

4. $\quad I_4 = \displaystyle\int x \ln\left(1 + \frac{1}{x}\right) dx.$

5. $\quad I_5 = \displaystyle\int_0^1 \sin^2(\ln x) dx.$

6. $\quad I_6 = \displaystyle\int \frac{dx}{1 + 3e^x}.$

7   $I_7 = \displaystyle\int_{\pi/4}^{\pi/3} \frac{dx}{\sin^3 x \cos^5 x}$.

8   $I_8 = \displaystyle\int_1^\infty \frac{dx}{x\sqrt{x^4 - 1}}$.

9   $I_9 = \displaystyle\int \frac{dx}{x(x^5 + 1)}$.

10   $I_{10} = \displaystyle\int_0^{\pi/4} \sqrt{\tan x}\, dx$.

11   $I_{11} = \displaystyle\int_0^1 \frac{\ln(1 + x)}{1 + x^2} dx$.

12   $I_{12} = \displaystyle\int_{64}^{729} \frac{\sqrt{x}}{\sqrt{x} - \sqrt[3]{x}} dx$.

13   $I_{13} = \displaystyle\int x^x(1 + \ln x)\, dx$.

14   $I_{14} = \displaystyle\int_0^1 x^{13/2}\sqrt{1 + x^{5/2}}\, dx$.

15   $I_{15} = \displaystyle\int_0^\infty \frac{dx}{(1 + x^2)^2}$.

16   $I_{16} = \displaystyle\int_0^1 \frac{dx}{x^4 - 13x^2 + 36}$.

17   $I_{17} = \displaystyle\int \frac{\ln(\ln x)}{x} dx$.

18   $I_{18} = \displaystyle\int \frac{1 + \cot x}{1 - \cot x} dx$.

$\boxed{19}$  $I_{19} = \displaystyle\int \dfrac{\cos x + x \sin x}{x(x + \cos x)}\,dx.$

$\boxed{20}$  $I_{20} = \displaystyle\int_0^{\pi/2} \dfrac{dx}{\sin x + \sec x}.$

$\boxed{21}$  $I_{21} = \displaystyle\int_0^{\infty} \dfrac{dx}{\sqrt{1 + e^x + e^{2x}}}.$

$\boxed{22}$  $I_{22} = \displaystyle\int_0^1 x^3 e^{x^2}\,dx.$

$\boxed{23}$  $I_{23} = \displaystyle\int_0^1 \sqrt{1 + x\sqrt{1 + x\sqrt{1 + x\sqrt{\ldots}}}}\,dx.$

$\boxed{24}$  $I_{24} = \displaystyle\int \left( \dfrac{1}{\ln x} - \dfrac{1}{\ln^2 x} \right) dx.$

$\boxed{25}$  $I_{25} = \displaystyle\int_1^2 (x - 1)^{1/2}(2 - x)^{1/2}\,dx.$

## 2.2  2011 MIT Integration Bee, Qualifying Test.

$\boxed{1}$  $I_1 = \displaystyle\int \dfrac{x^6 - 1}{x^4 + x^3 - x - 1}\,dx.$

$\boxed{2}$  $I_2 = \displaystyle\int \left( 2\ln x + (\ln x)^2 \right) dx.$

$\boxed{3}$  $I_3 = \displaystyle\int \dfrac{2x}{\sqrt{1 - x^4}}\,dx.$

$\boxed{4}$  $I_4 = \displaystyle\int \dfrac{x^2 + 1}{x + 1}\,dx.$

5  $I_5 = \displaystyle\int \frac{\sin^3 x + \sin^2 x - 2\sin x - 2}{\sin^2 x + 2\sin x + 1}dx.$

6  $I_6 = \displaystyle\int \sinh^{-2}(x)dx.$

7  $I_7 = \displaystyle\int \sec^4 x \tan^2 x dx.$

8  $I_8 = \displaystyle\int \sqrt{\csc x - \sin x}dx.$

9  $I_9 = \displaystyle\int \cos^6 x dx.$

10  $I_{10} = \displaystyle\int \frac{dx}{x^4 + 2x^2 + 1}.$

11  $I_{11} = \displaystyle\int \cos(\ln x)dx.$

12  $I_{12} = \displaystyle\int \frac{dx}{\cos x}.$

13  $I_{13} = \displaystyle\int \frac{dx}{9\cos^2 x + 4\sin^2 x}.$

14  $I_{14} = \displaystyle\int \frac{dx}{x^2(x^4 + 1)^{3/4}}.$

15  $I_{15} = \displaystyle\int_0^\pi \cos x \cos(3x) \cos(5x)dx.$

16  $I_{16} = \displaystyle\int \left( \frac{1}{\ln x} + \ln(\ln x) \right) dx.$

17    $I_{17} = \displaystyle\int \frac{dx}{2 + e^x}.$

18    $.I_{18} = \displaystyle\int \sqrt{\frac{x}{1 - x^3}}\, dx.$

19    $I_{19} = \displaystyle\int \frac{4x}{1 - x^4}\, dx.$

20    $I_{20} = \displaystyle\int x^x(1 + \ln x)\, dx.$

21    $I_{21} = \displaystyle\int_0^6 \sqrt{6x - x^2}\, dx.$

22    $I_{22} = \displaystyle\int \sin(101x) \sin^{99} x\, dx.$

23    $I_{23} = \displaystyle\int x\, e^{e^{x^2} + x^2}\, dx.$

24    $I_{24} = \displaystyle\int_0^1 \frac{x^3 - 3x^2 + 3x - 1}{x^4 + 4x^3 + 6x^2 + 4x + 1}\, dx.$

25    $I_{25} = \displaystyle\int \sqrt{\frac{1 - x}{1 + x}}\, dx.$

## 2.3   2012 MIT Integration Bee, Qualifying Test.

1    $I_1 = \displaystyle\int \frac{dx}{\sqrt{x} - 1}.$

2    $I_2 = \displaystyle\int x^{1/4} \ln x\, dx.$

3   $$I_3 = \int \frac{dx}{(1 + \sqrt{x})\sqrt{x - x^2}}.$$

4   $$I_4 = \int \frac{dx}{\sqrt{x}\left(\sqrt[4]{x} + 1\right)^{10}}.$$

5   $$I_5 = \int_0^1 \sin\left(\cos^{-1} x\right) dx.$$

6   $$I_6 = \int \frac{dx}{\sqrt{1 - 4x - x^2}}.$$

7   $$I_7 = \int_{1/4}^{1/2} \left\lfloor \ln \left\lfloor \frac{1}{x} \right\rfloor \right\rfloor dx.$$

8   $$I_8 = \int_0^{\pi/2} \frac{dx}{1 + \sin x}.$$

9   $$I_9 = \int_1^{2011} \frac{\sqrt{x}}{\sqrt{2012 - x} + \sqrt{x}} dx.$$

10   $$I_{10} = \int \frac{x - 1}{(x + 1)\sqrt{x^3 + x^2 + x}} dx.$$

11   $$I_{11} = \int_{-1}^0 \frac{x^4 + 4x^3 + 6x^2 + 4x + 1}{x^3 - 3x^2 + 3x - 1} dx.$$

12   $$I_{12} = \int \left(\cos x \ln x + \frac{\sin x}{x}\right) dx.$$

13   $$I_{13} = \int \frac{dx}{x^3 - x}.$$

14   $$I_{14} = \int_0^{1/2} \frac{x \sin^{-1} x}{\sqrt{1 - x^2}} dx.$$

15  $I_{15} = \displaystyle\int_0^1 x(1-x)^{99} dx.$

16  $I_{16} = \displaystyle\int_0^{\pi/2} \frac{\sin(4x)}{\sin x} dx.$

17  $I_{17} = \displaystyle\int \frac{x^{-1/2}}{1 + x^{1/3}} dx.$

18  $I_{18} = \displaystyle\int \frac{dx}{\sqrt{2x^2 - 1}}.$

19  $I_{19} = \displaystyle\int \frac{dx}{\sqrt{e^x - 1}}.$

20  $I_{20} = \displaystyle\int \frac{x}{x^4 + 4} dx.$

21  $I_{21} = \displaystyle\int \frac{2dx}{(\cos x - \sin x)^2}.$

22  $I_{22} = \displaystyle\int \frac{x \cosh x}{\sinh^2 x} dx.$

23  $I_{23} = \displaystyle\int_0^2 x^5 \sqrt{1 + x^3} dx.$

24  $I_{24} = \displaystyle\int_0^1 \frac{x^7 - 1}{\ln x} dx.$

25  $I_{25} = \displaystyle\int \sqrt{\csc x - \sin x} dx.$

## 2.4    2013 MIT Integration Bee, Qualifying Test.

$\boxed{1}$   $I_1 = \displaystyle\int \left(\ln(x^2) - 2\ln(2x)\right)\,dx.$

$\boxed{2}$   $I_2 = \displaystyle\int_{-1}^{3} e^{|x|}\,dx.$

$\boxed{3}$   $I_3 = \displaystyle\int \frac{\ln x \cos x - (\sin x/x)}{\ln^2 x}\,dx.$

$\boxed{4}$   $I_4 = \displaystyle\int_{1}^{11} \left(x^3 - 3x^2 + 3x - 1\right)\,dx.$

$\boxed{5}$   $I_5 = \displaystyle\int_{0}^{2} \sqrt{12 - 3x^2}\,dx.$

$\boxed{6}$   $I_6 = \displaystyle\int_{0}^{6} \left(x + (x - 3)^7 + \sin(x - 3)\right)\,dx.$

$\boxed{7}$   $I_7 = \displaystyle\int \sin x\sqrt{1 + \tan^2 x}\,dx.$

$\boxed{8}$   $I_8 = \displaystyle\int \frac{x^5 - x^3 + x^2 - 1}{x^4 - x^3 + x - 1}\,dx.$

$\boxed{9}$   $I_9 = \displaystyle\int_{0}^{1} \ln x\,dx.$

$\boxed{10}$   $I_{10} = \displaystyle\int \frac{dx}{1 - e^{-x}}.$

$\boxed{11}$   $I_{11} = \displaystyle\int_{0}^{\pi} \sin^2 x \cos^2 x\,dx.$

12    $I_{12} = \displaystyle\int_0^{441} \frac{\pi \sin\left(\pi\sqrt{x}\right)}{\sqrt{x}}\, dx.$

13    $I_{13} = \displaystyle\int \tan^2 x\, dx.$

14    $I_{14} = \displaystyle\int_0^{256} \left(x - \lfloor x \rfloor\right)^2 dx.$

15    $I_{15} = \displaystyle\int e^{\sqrt[4]{x}} dx.$

16    $I_{16} = \displaystyle\int \cos x \cot x\, dx.$

17    $I_{17} = \displaystyle\int \left(2\ln x + \ln^2 x\right) dx.$

18    $I_{18} = \displaystyle\int \frac{x^3}{1 + x^2}\, dx.$

19    $I_{19} = \displaystyle\int \frac{dx}{2 - 2x + x^2}.$

20    $I_{20} = \displaystyle\int \sin x \ln\left(\sin x\right) dx.$

21    $I_{21} = \displaystyle\int \frac{x}{1 - x^4}\, dx.$

22    $I_{22} = \displaystyle\int \sqrt{12 - 3x^2}\, dx.$

23    $I_{23} = \displaystyle\int \sec^5 x \tan^3 x\, dx.$

$$\boxed{24} \quad I_{24} = \int_{-\pi/4}^{\pi/4} \frac{dx}{1 - \sin x}.$$

$$\boxed{25} \quad I_{25} = \int \frac{dx}{x\sqrt{x^2 - 2}}.$$

## 2.5  2014 MIT Integration Bee, Qualifying Test.

$$\boxed{1} \quad I_1 = \int_1^e \ln(x^2)dx.$$

$$\boxed{2} \quad I_2 = \int_{-9}^9 \sin\left(\sqrt[3]{x}\right) dx.$$

$$\boxed{3} \quad I_3 = \int_0^\infty \left(\frac{d}{dx}\left(e^{1+x-x^2}\right)\right) dx.$$

$$\boxed{4} \quad I_4 = \int_0^2 \sqrt{x + \sqrt{x + \sqrt{x + \sqrt{x + \ldots}}}}\,dx.$$

$$\boxed{5} \quad I_5 = \int \sqrt{x}\, e^{\sqrt{x}}dx.$$

$$\boxed{6} \quad I_6 = \int \sin(2x)\cos(3x)dx.$$

$$\boxed{7} \quad I_7 = \int_0^{2\pi} |1 + 2\sin x|\, dx.$$

$$\boxed{8} \quad I_8 = \int x(1 - x)^{2014}dx.$$

$$\boxed{9} \quad I_9 = \int \operatorname{arsinh} x\, dx.$$

**10** $I_{10} = \int_{-1}^{0} \frac{x^2}{x - 1} dx.$

**11** $I_{11} = \int x \arctan x \, dx.$

**12** $I_{12} = \int \frac{dx}{x^2 - 15x - 2014}.$

**13** $I_{13} = \int e^x \left( \ln(1 + x^2) - 2(1 + x) \arctan x \right) dx.$

**14** $I_{14} = \int (\arcsin x)^2 \, dx.$

**15** $I_{15} = \int \frac{\sqrt{x^2 - 1}}{x} dx.$

**16** $I_{16} = \int x \sec^2(4x) dx.$

**17** $I_{17} = \int \frac{2}{6 - 11x + 6x^2 - x^3} dx.$

**18** $I_{18} = \int_0^1 \frac{dx}{\lfloor 1 - \log_2(1 - x) \rfloor}.$

**19** $I_{19} = \int_0^{1/\sqrt{3}} \sqrt{x + \sqrt{x^2 + 1}} \, dx.$

**20** $I_{20} = \int_0^{5\pi/2} \frac{dx}{2 + \cos x}.$

## 2.6   2015 MIT Integration Bee, Qualifying Test.

$\boxed{1}$ $I_1 = \displaystyle\int \left( \cos^4 x - \sin^4 x \right) dx.$

$\boxed{2}$ $I_2 = \displaystyle\int \frac{x}{\sqrt{2 + 4x}} dx.$

$\boxed{3}$ $I_3 = \displaystyle\int_0^8 \frac{\cos\left(\sqrt{x}\right)}{\sqrt{x}} dx.$

$\boxed{4}$ $I_4 = \displaystyle\int \sec x \, dx.$

$\boxed{5}$ $I_5 = \displaystyle\int_0^{\pi/2} \frac{e^{\sin x}}{\tan x \csc x} dx.$

$\boxed{6}$ $I_6 = \displaystyle\int_1^e x \ln^2 x \, dx.$

$\boxed{7}$ $I_7 = \displaystyle\int \frac{dx}{5 + 4\sqrt{x} + x}.$

$\boxed{8}$ $I_8 = \displaystyle\int (2015)^x dx.$

$\boxed{9}$ $I_9 = \displaystyle\int_0^2 \frac{x}{(x-3)(x+5)^2} dx.$

$\boxed{10}$ $I_{10} = \displaystyle\int \frac{\ln\left(1 + \ln x\right)}{x} dx.$

$\boxed{11}$ $I_{11} = \displaystyle\int \sqrt{\csc x - \sin x} \, dx.$

12  $I_{12} = \displaystyle\int \frac{dx}{\sqrt{x^2 + 25}}.$

13  $I_{13} = \displaystyle\int_2^e \frac{\ln^2 x - 1}{x \ln^2 x}\, dx.$

14  $I_{14} = \displaystyle\int e^{3x} \arctan\left(e^x\right) dx.$

15  $I_{15} = \displaystyle\int_0^4 \frac{|x-1|}{|x-2| + |x-3|}\, dx.$

16  $I_{16} = \displaystyle\int_0^{2\pi} \frac{dx}{\sin^4 x + \cos^4 x}.$

17  $I_{17} = \displaystyle\int \frac{1 + e^x}{1 - e^x}\, dx.$

18  $I_{18} = \displaystyle\int \tan^4 x\, dx.$

19  $I_{19} = \displaystyle\int \sin x \tan^2 x\, dx.$

20  $I_{20} = \displaystyle\int \frac{x+1}{x^2 + 2x + 3}\, dx.$

## 2.7   2016 MIT Integration Bee, Qualifying Test.

1  $I_1 = \displaystyle\int \tanh x\, dx.$

2  $I_2 = \displaystyle\int_{-4}^4 \left| x^3 - x \right| dx.$

3  $\quad I_3 = \displaystyle\int_1^e \ln\left(\sqrt{x}\right) dx.$

4  $\quad I_4 = \displaystyle\int \left(e^{e^x + e^{-x} + x} - e^{e^x + e^{-x} - x}\right) dx.$

5  $\quad I_5 = \displaystyle\int \frac{\ln\left(\ln x\right)}{x \ln x} dx.$

6  $\quad I_6 = \displaystyle\int_0^{\pi/3} \frac{dx}{1 + \tan^2 x}.$

7  $\quad I_7 = \displaystyle\int_{-27}^{27} \arcsin\left(\sqrt[3]{x}/3\right) dx.$

8  $\quad I_8 = \displaystyle\int_{50}^{100} \lfloor \log_2 x \rfloor \, dx.$

9  $\quad I_9 = \displaystyle\int \left(e^x \cos x - e^x \sin x\right) dx.$

10  $\quad I_{10} = \displaystyle\int_0^\infty x^3 e^{-x^2} dx.$

11  $\quad I_{11} = \displaystyle\int \left(\left(2xe^{x^2} + 1\right) \cos x - \left(e^{x^2} + x\right) \sin x\right) dx.$

12  $\quad I_{12} = \displaystyle\int \left(1 + x^{1/2} + x^{1/3}\right) \left(1 + x^{-1/2} + x^{-1/3}\right) dx.$

13  $\quad I_{13} = \displaystyle\int \sin\left(\sin x\right) \cos\left(\sin x\right) \cos x \, dx.$

14  $\quad I_{14} = \displaystyle\int \left(\frac{\cos x + \sin x}{x^2} + \frac{\sin x - \cos x}{x}\right) dx.$

15  $I_{15} = \int x^3 \sqrt{x^2 + 1}\, dx.$

16  $I_{16} = \int \dfrac{x}{x^4 + x^2 + 1}\, dx.$

17  $I_{17} = \int e^{e^{2016\, x} + 6048x}\, dx.$

18  $I_{18} = \int_{\pi/3}^{\pi/2} \dfrac{1 - \cos x}{\sin x}\, dx.$

19  $I_{19} = \int \dfrac{dx}{1 - x + x^2 - x^3}.$

20  $I_{20} = \int_0^{\infty} \dfrac{dx}{2 + \cosh x}.$

## 2.8  2017 MIT Integration Bee, Qualifying Test.

1  $I_1 = \int \dfrac{x^2}{\sqrt{x^3 + 2}}\, dx.$

2  $I_2 = \int_1^{\infty} \dfrac{\ln x}{x^2}\, dx.$

3  $I_3 = \int \operatorname{sech} x\, dx.$

4  $I_4 = \int x^3 e^{x^2}\, dx.$

5  $I_5 = \int_1^2 \dfrac{dx}{x\sqrt{x^2 - 1}}.$

6   $I_6 = \int_1^\infty \dfrac{dx}{x(x^2+1)}.$

7   $I_7 = \int \cosh^{-1} x\, dx.$

8   $I_8 = \int_{-\infty}^\infty e^{-2x^2-5x-3}\, dx.$

9   $I_9 = \int \sin\left(\sqrt{x}\right)\, dx.$

10   $I_{10} = \int_0^\infty \dfrac{dx}{\left(x+\frac{1}{x}\right)^2}.$

11   $I_{11} = \int \dfrac{(2+x)e^{-x}}{x^3}\, dx.$

12   $I_{12} = \int_0^1 \dfrac{dx}{\sqrt{x(1-x)}}.$

13   $I_{13} = \int_0^\infty \dfrac{\tanh x}{e^x}\, dx.$

14   $I_{14} = \int_0^{\pi/2} \sqrt{1+\sin x}\, dx.$

15   $\lim\limits_{n\to\infty} I_n = ?$   where   $I_1 = \int_0^1 \dfrac{dx}{1+\sqrt{x}}, \quad I_2 = \int_0^1 \dfrac{dx}{1+\frac{1}{1+\sqrt{x}}}, \quad \cdots$

16   $I_{16} = \int_{-\infty}^\infty \dfrac{\sin^2(x+\pi/4)}{e^{x^2}}\, dx.$

17   $I_{17} = \int_{-\infty}^\infty 3x^2\left(x^3+1\right)^2 e^{-x^6-2x^3}\, dx.$

18 $\quad I_{18} = \int_0^{\pi/2} \dfrac{dx}{1 + \tan^{2017} x}.$

19 $\quad I_{19} = \int e^{2x} \cos(3x)dx.$

20 $\quad I_{20} = \int \left((\cos x)^{\cos x + 1} \tan x\right)(1 + \ln(\cos x))dx.$

## 2.9  2018 MIT Integration Bee, Qualifying Test.

1 $\quad I_1 = \int \dfrac{e^x}{e^x + 2}dx.$

2 $\quad I_2 = \int \sqrt{x \sqrt[3]{x \sqrt[4]{x \sqrt[5]{x \dots}}}}\,dx.$

3 $\quad I_3 = \int_0^{2018\pi} |\sin(2018x)|\,dx.$

4 $\quad I_4 = \int \dfrac{dx}{\tan x + \cot x}.$

5 $\quad I_5 = \int \dfrac{x^5}{2 + x^{12}}dx.$

6 $\quad I_6 = \int (\cos x \cosh x + \sin x \sinh x)\,dx.$

7 $\quad I_7 = \int \dfrac{e^x + \cos x}{e^x + \sin x}dx.$

8 $\quad I_8 = \int \sin(\cos(\sin x)) \sin(\sin x) \cos x\,dx.$

9    $I_9 = \displaystyle\int \frac{dx}{1 + \sin x}.$

10    $I_{10} = \displaystyle\int \frac{\cos x}{1 - \cos(2x)} dx.$

11    $I_{11} = \displaystyle\int e^x \left( \frac{1}{x} + \ln x \right) dx.$

12    $I_{12} = \displaystyle\int \tanh^2 x\, dx.$

13    $I_{13} = \displaystyle\int \frac{2017 x^{2016} + 2018 x^{2017}}{1 + x^{4034} + 2 x^{4035} + x^{4036}} dx.$

14    $I_{14} = \displaystyle\int \frac{\sin(2x) - \sin^2 x}{\cos(2x) - \cos^2 x} dx.$

15    $I_{15} = \displaystyle\int \frac{dx}{x^{25/25} x^{16/25} + x^{9/25}}.$

16    $I_{16} = \displaystyle\int_0^{\pi/2} \frac{\cos x}{2 - \cos^2 x} dx.$

17    $I_{17} = \displaystyle\int \frac{dx}{(1 + x^2)^{3/2}}.$

18    $I_{18} = \displaystyle\int \frac{dx}{\sqrt{x\sqrt{x} - x^2}}.$

19    $I_{19} = \displaystyle\int \frac{x - 1}{x + x^2 \ln x} dx.$

20    $I_{20} = \displaystyle\int \csc x \sec x\, dx.$

## 2.10   2019 MIT Integration Bee, Qualifying Test.

1  $I_1 = \displaystyle\int_0^{2\pi} \tan(\cos x)\,dx.$

2  $I_2 = \displaystyle\int \frac{x+1}{x(x+\ln x)}\,dx.$

3  $I_3 = \displaystyle\int \left(e^{x+e^x} + e^{x-e^x}\right)\,dx.$

4  $I_4 = \displaystyle\int_{-1/2}^{1/2} \frac{dx}{1-x^2}.$

5  $I_5 = \displaystyle\int_0^2 2^{\ln x}\,dx.$

6  $I_6 = \displaystyle\int_{-2\pi}^{2\pi} \left(\cos(3x)+\sin(2x)\right)\left(-\sin(2019x)+\cos(3x)\right)\,dx.$

7  $I_7 = \displaystyle\int \cos x\,\left(\cos(\sin x)\right)\cos\left(\sin(\sin x)\right)\,dx.$

8  $I_8 = \displaystyle\int_0^\infty \frac{e^{-2019/(4t^2)}}{t^2}\,dt.$

9  $I_9 = \displaystyle\int \sin\left(\sqrt{x}\right)\,dx.$

10  $I_{10} = \displaystyle\int_0^1 \frac{\sqrt{x}}{1+x}\,dx.$

11  $I_{11} = \displaystyle\int_0^{2\pi} \cos x\,\cos(2x)\,\cos(3x)\,dx.$

**12**  $I_{12} = \lim\limits_{n \to \infty} \int_{-\infty}^{\infty} e^{-x^{2n}}\, dx.$

**13**  $I_{13} = \int_{0}^{e} x^{1/\ln x}\, dx.$

**14**  $I_{14} = \int_{0}^{\pi/100} \dfrac{\sin(20x) + \sin(19x)}{\cos(20x) + \cos(19x)}\, dx.$

**15**  $I_{15} = \int \left( e^x \cos^2 x + e^x \sin x \cos x - e^x \sin^2 x \right) dx.$

**16**  $I_{16} = \int_{0}^{\pi/2} \dfrac{\sin x}{\sin\left(x + \frac{\pi}{4}\right)}\, dx.$

**17**  $I_{17} = \int \dfrac{dx}{x + \sqrt[3]{x}}.$

**18**  $I_{18} = \int_{0}^{2} x^{x^2+1} \left(2\ln x + 1\right) dx.$

**19**  $I_{19} = \int \dfrac{2x^3 - 1}{x(x^3 + 1)}\, dx.$

**20**  $I_{20} = \int \cos\left(\arctan x\right) dx.$

## 2.11  2020 MIT Integration Bee, Qualifying Test.

**1**  $I_1 = \int \dfrac{\ln(2x)}{x \ln x}\, dx.$

**2**  $I_2 = \int_{0}^{\infty} \dfrac{dx}{e^x + 1}.$

$\boxed{3}$   $I_3 = \displaystyle\int_e^{e^e} \frac{\ln x \ln (\ln x)}{x} dx.$

$\boxed{4}$   $I_4 = \displaystyle\int_0^1 \ln \left(\frac{1+x}{1-x}\right) dx.$

$\boxed{5}$   $I_5 = \displaystyle\int \frac{dx}{x^2 + (x-1)^2}.$

$\boxed{6}$   $I_6 = \displaystyle\int \sqrt{x\sqrt{x\sqrt{x\sqrt{x\ldots}}}}\, dx.$

$\boxed{7}$   $I_7 = \displaystyle\int \sin^4 x \cos^4 x (\cos x + \sin x)(\cos x - \sin x) dx.$

$\boxed{8}$   $I_8 = \displaystyle\int \ln(1 + x^2) dx.$

$\boxed{9}$   $I_9 = \displaystyle\int_0^{2\pi} (\cos x)^{2020} dx.$

$\boxed{10}$   $I_{10} = \displaystyle\int \frac{2x+1}{2x^2 + 2x + 1} dx.$

$\boxed{11}$   $I_{11} = \displaystyle\int_{1/\sqrt{2}}^1 \frac{\arcsin x}{x^3} dx.$

$\boxed{12}$   $I_{12} = \displaystyle\int_0^{\pi/2} \sin(2x) \cos(\cos x) dx.$

$\boxed{13}$   $I_{13} = \displaystyle\int_0^{2\pi} \sin (\sin x - x) \, dx.$

$\boxed{14}$   $I_{14} = \displaystyle\int \left(\frac{1}{x-1} + \frac{\sum_{k=0}^{2018} (k+1)x^k}{\sum_{k=0}^{2019} x^k}\right) dx.$

15 $\quad I_{15} = \displaystyle\int_0^{\pi/2} \dfrac{dx}{\tan^{\sqrt{2020}} x + 1}.$

16 $\quad I_{16} = \displaystyle\int x(1-x)^{2020} dx.$

17 $\quad I_{17} = \displaystyle\int \dfrac{\sec^4 x \tan x}{\sec^4 x + 4} dx.$

18 $\quad I_{18} = \displaystyle\int x^{2x} \left(2 + 2\ln x\right) dx.$

19 $\quad I_{19} = \displaystyle\int_0^1 \sqrt{1 - x^2} dx.$

20 $\quad I_{20} = \displaystyle\int_0^{\infty} x^5 e^{-x^4} dx.$

## 2.12    2022 MIT Integration Bee, Qualifying Test.

1 $\quad I_1 = \displaystyle\int \dfrac{1 + \cos x}{x + \sin x} dx.$

2 $\quad I_2 = \displaystyle\int_1^{\sqrt{3}} \dfrac{\arctan x + \operatorname{arccot} x}{x} dx.$

3 $\quad I_3 = \displaystyle\int_0^{2022} \left(x^2 - \lfloor x \rfloor \lceil x \rceil\right) dx.$

4 $\quad I_4 = \displaystyle\int \dfrac{\sinh x}{\cosh x - \sinh x} dx.$

5 $\quad I_5 = \displaystyle\int \dfrac{x}{\sqrt{x-1} + \sqrt{x+1}} dx.$

6  $I_6 = \int_0^\pi \cos(x + \cos x)dx.$

7  $I_7 = \int x^3 \sin(x^2)dx.$

8  $I_8 = \int \frac{x}{1 - x^4}dx.$

9  $I_9 = \int \frac{dx}{\cosh^2 x}.$

10  $I_{10} = \int_0^1 \left(e^{e^x} - e^{e^x - x}\right) dx.$

11  $I_{11} = \lim_{n \to \infty} \int_0^3 \underbrace{\sin\left(\frac{\pi}{3}\sin\left(\frac{\pi}{3}\sin\left(\ldots\sin\left(\frac{\pi}{3}x\right)\right)\right)\right)}_{n \text{ sin's}} dx.$

12  $I_{12} = \int_0^1 \sqrt{1 - \sqrt{x}}dx.$

13  $I_{13} = \int \frac{x^3}{1 + x + \frac{x^2}{2} + \frac{x^3}{6}}dx.$

14  $I_{14} = \int (\sin(x + \sin x) - \sin(x - \sin x)) \, dx.$

15  $I_{15} = \int \left(\tan^4 x \sec^3 x + \tan^2 x \sec^5 x\right) dx.$

16  $I_{16} = \int (1 + \ln x) \ln(\ln x) \, dx.$

$\boxed{17}$   $I_{17} = \int \left( \dfrac{1}{1+\sin x} + \dfrac{1}{1+\cos x} + \dfrac{1}{1+\tan x} + \dfrac{1}{1+\cot x} \right.$
$$\left. + \dfrac{1}{1+\sec x} + \dfrac{1}{1+\csc x} \right) dx.$$

$\boxed{18}$   $I_{18} = \displaystyle\int \dfrac{dx}{\sqrt{x - x^2}}.$

$\boxed{19}$   $I_{19} = \displaystyle\int_0^{1/2} \left( \sum_{n=0}^{\infty} \binom{n+3}{n} x^n \right) dx.$

$\boxed{20}$   $I_{20} = \displaystyle\int \dfrac{dx}{1 + \cos^2 x}.$

## 2.13   2023 MIT Integration Bee, Qualifying Test.

$\boxed{1}$   $I_1 = \displaystyle\int x^{1/\ln x}\, dx.$

$\boxed{2}$   $I_2 = \displaystyle\int \operatorname{sech} x\, dx.$

$\boxed{3}$   $I_3 = \displaystyle\int \dfrac{e^x}{(1+e^x)\ln(1+e^x)}\, dx.$

$\boxed{4}$   $I_4 = \displaystyle\int \left(1 + x + x^2 + x^3 + x^4\right)\left(1 - x + x^2 - x^3 + x^4\right) dx.$

$\boxed{5}$   $I_5 = \displaystyle\int_0^4 \left( \dfrac{x}{5} \right) dx.$

$\boxed{6}$   $I_6 = \displaystyle\int (x + \sin x + x\cos x + \sin x \cos x)\, dx.$

$\boxed{7}$   $I_7 = \displaystyle\int \left( \sin^2 x + \cos^2 x + \tan^2 x + \cot^2 x + \sec^2 x + \csc^2 x \right) dx.$

8  $I_8 = \displaystyle\int_0^{2\pi} \lfloor 2023 \sin x \rfloor dx.$

9  $I_9 = \displaystyle\int (1 + 2\ln x)\, e^{(\ln x)^2}\, dx.$

10  $I_{10} = \displaystyle\int \Big( (1-x)^3 + \left(x - x^2\right)^3 + \left(x^2 - 1\right)^3$
$$- 3(1-x)\left(x - x^2\right)\left(x^2 - 1\right) \Big) dx.$$

11  $I_{11} = \displaystyle\int_{-2023}^{2023} \underbrace{||||\,|x| - 1| - 1|\ldots| - 1|}_{2023(-1)'s}\, dx.$

12  $I_{12} = \displaystyle\int \left(\sin^6 x + \cos^6 x + 3\sin^2 x \cos^2 x\right) dx.$

13  $I_{13} = \displaystyle\int (x + e + 1)x^e e^x dx.$

14  $I_{14} = \displaystyle\int_0^1 \left(\frac{x^2}{2 - x^2} + \sqrt{\frac{2x}{x+1}}\right) dx.$

15  $I_{15} = \displaystyle\int \frac{1 + 2x^{2022}}{x + x^{2023}}\, dx.$

16  $I_{16} = \displaystyle\int \left(3\sin(20x)\cos(23x) + 20\sin(43x)\right) dx.$

17  $I_{17} = \displaystyle\int_0^1 \prod_{k=0}^{\infty} \left(\frac{1}{1 + x^{2^k}}\right) dx.$

18  $I_{18} = \displaystyle\int \frac{\sin x}{2e^x + \cos x + \sin x}\, dx.$

19  $I_{19} = \displaystyle\int \frac{\ln(x/\pi)}{(\ln x)^{\ln(e\pi)}}\,dx.$

20  $I_{20} = \displaystyle\int_{-3/2}^{-1/2} \left(x^5 + 5x^4 + 10x^3 + 8x^2 + x\right) dx.$

# Chapter 3

# The Solutions to the 2010 MIT Integration Bee, Qualifying Test

$$\boxed{I_1 = \int_0^{\pi/2} \sin x \sin(2x) \sin(3x) dx}$$

**Solution.**

For the given integral we can write the following

$$I_1 = -\frac{1}{2} \int_0^{\pi/2} (\cos(3x) - \cos x) \sin(3x) dx$$

$$= -\frac{1}{2} \int_0^{\pi/2} (\sin(3x)\cos(3x) - \sin(3x)\cos x) \, dx$$

$$= -\frac{1}{2} \int_0^{\pi/2} \left( \frac{1}{2}\sin(6x) - \frac{1}{2}(\sin(4x) + \sin(2x)) \right) dx$$

$$= -\frac{1}{4} \int_0^{\pi/2} (\sin(6x) - \sin(4x) - \sin(2x)) \, dx$$

$$= -\frac{1}{4} \left[ -\frac{1}{6}\cos(6x) + \frac{1}{4}\cos(4x) + \frac{1}{2}\cos(2x) \right]_0^{\pi/2}$$

$$= -\frac{1}{4} \left( -\frac{2}{3} \right) = \frac{1}{6}.$$

$$\boxed{I_2 = \int_0^{\pi/2} \sin^3(2x) \cos x \, dx}$$

**Solution.**

For the given integral we can write the following

$$
I_2 = \int_0^{\pi/2} (2\sin x \cos x)^3 \cos x dx = 8 \int_0^{\pi/2} \sin^3 x \cos^4 x dx
$$

$$
= 8 \int_0^{\pi/2} \cos^4 x \sin^2 x \sin x dx = 8 \int_0^{\pi/2} \cos^4 x (1 - \cos^2 x) \sin x dx.
$$

For the last integral, let us use the substitution $\cos x = t$, so that $\sin x dx = -dt$. If $x = 0$, then $t = 1$, and if $x = \pi/2$, then $t = 0$. With these new values, we get the following

$$
I_2 = -8 \int_1^0 t^4 (1 - t^2) dt = 8 \int_0^1 (t^4 - t^6) dt = 8 \left[ \frac{t^5}{5} - \frac{t^7}{7} \right]_0^1
$$

$$
= 8 \left( \frac{1}{5} - \frac{1}{7} \right) = \frac{16}{35}.
$$

$$
\boxed{ I_3 = \int (x+1)^2 (x-1)^{1/3} dx }
$$

**Solution 1.**

For the given integral we can write the following

$$
I_3 = \int \left( x^2 (x-1)^{1/3} + 2x(x-1)^{1/3} + (x-1)^{1/3} \right) dx
$$

$$
= \int (x^2 - 1 + 1)(x-1)^{1/3} dx + 2 \int (x - 1 + 1)(x-1)^{1/3} dx
$$

$$
+ \int (x-1)^{1/3} dx
$$

$$
= \int (x-1)(x+1)(x-1)^{\frac{1}{3}} dx + \int (x-1)^{\frac{1}{3}} dx + 2 \int (x-1)(x-1)^{\frac{1}{3}} dx
$$

$$
+ 2 \int (x-1)^{\frac{1}{3}} dx + \int (x-1)^{\frac{1}{3}} dx
$$

$$
= \int (x - 1 + 2)(x-1)^{4/3} dx + 2 \int (x-1)^{4/3} dx + 4 \int (x-1)^{1/3} dx
$$

$$
= \int (x-1)^{7/3} dx + 4 \int (x-1)^{4/3} dx + 4 \int (x-1)^{1/3} dx
$$

$$
= \frac{3}{10} (x-1)^{10/3} + \frac{12}{7} (x-1)^{7/3} + 3(x-1)^{4/3} + C
$$

$$
= 3(x-1)^{4/3} \left( \frac{1}{10} (x-1)^2 + \frac{4}{7} (x-1) + 1 \right) + C,
$$

where $C \in \mathbb{R}$.

**Solution 2.**

By using the integration by parts (tabular integration), we find

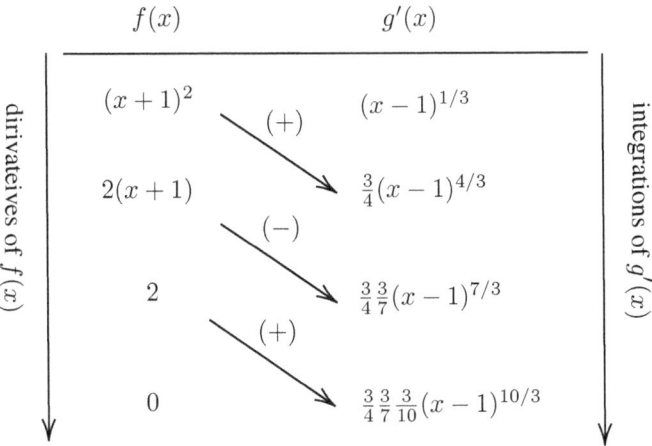

$$I_3 = \frac{3}{4}(x+1)^2(x-1)^{4/3} - \frac{9}{14}(x+1)(x-1)^{7/3} + \frac{27}{140}(x-1)^{10/3} + C,$$

where $C \in \mathbb{R}$.

$$\boxed{I_4 = \int x \ln\left(1 + \frac{1}{x}\right) dx}$$

**Solution.**

At the first, let us assume that $x > 0$ (Be attentive when the integral is definite!).

By using the integration by parts, let us assume

$$u = \ln\left(1 + \frac{1}{x}\right) \implies du = -\frac{dx}{x(x+1)}, \quad dv = x dx \implies v = \frac{x^2}{2}.$$

Thus, we have

$$I_6 = \frac{x^2}{2} \ln\left(1 + \frac{1}{x}\right) + \frac{1}{2} \int \frac{x^2}{x(x+1)} dx$$

$$= \frac{x^2}{2} \ln\left(1 + \frac{1}{x}\right) + \frac{1}{2} \int \left(1 - \frac{1}{1+x}\right) dx$$

$$= \frac{x^2}{2} \ln\left(1 + \frac{1}{x}\right) + \frac{1}{2}x - \frac{1}{2}\ln(1+x) + C,$$

where $C \in \mathbb{R}$.

$$I_5 = \int_0^1 \sin^2(\ln x)dx$$

**Solution.**

At the first, we note that the given integral is improper, where the integrand $f(x) = \sin^2(\ln x)$ is not continuous at $x = 0$. So, we will write

$$I_5 = \lim_{\varepsilon \to 0^+} \int_\varepsilon^1 \sin^2(\ln x)dx.$$

Let us calculate the indefinite integral

$$I = \int \sin^2(\ln x)dx. \tag{3.1}$$

For this integral, we can write

$$I = \int \frac{1 - \cos(2\ln x)}{2}dx = \frac{1}{2}\int dx - \frac{1}{2}\underbrace{\int \cos(2\ln x)dx}_{:=J}.$$

For the integral $J$, by using the integration by parts, let us assume

$$u = \cos(2\ln x) \Longrightarrow du = -\frac{2\sin(2\ln x)}{x}dx, \quad dv = dx \Longrightarrow v = x.$$

Thus, we have

$$J = x\cos(2\ln x) + 2\int \sin(2\ln x)dx.$$

Now, for the last integral, by using the integration by parts, let us assume

$$u = \sin(2\ln x) \Longrightarrow du = \frac{2\cos(2\ln x)}{x}dx, \quad dv = dx \Longrightarrow v = x.$$

Thus, we have

$$J = x\cos(2\ln x) + 2\left(x\sin(2\ln x) - 2\underbrace{\int \cos(2\ln x)dx}_{J}\right).$$

From this, we get

$$5J = x\cos(2\ln x) + 2x\sin(2\ln x).$$

Therefore

$$J = \frac{1}{5}x\cos(2\ln x) + \frac{2}{5}x\sin(2\ln x) + C,$$

where $C \in \mathbb{R}$. As a result, for the integral (3.1), we have

$$I = \frac{1}{2}x - \frac{1}{10}x\cos(2\ln x) - \frac{1}{5}x\sin(2\ln x) + C. \tag{3.2}$$

Therefore, for the given integral $I_5$, we get the following

$$I_5 = \lim_{\varepsilon \to 0^+} \left[ \frac{1}{2}x - \frac{1}{10}x\cos(2\ln x) - \frac{1}{5}x\sin(2\ln x) \right]_{\varepsilon^+}^{1}$$

$$= \frac{1}{2} - \frac{1}{10} - \lim_{\varepsilon \to 0} \left( \frac{\varepsilon}{2} - \frac{\varepsilon}{10}\cos(2\ln\varepsilon) - \frac{\varepsilon}{5}\sin(2\ln\varepsilon) \right)$$

$$= \frac{1}{2} - \frac{1}{10} = \frac{2}{5}.$$

**Remark.** For the integral $I$ in (3.1). Let us use the substitution $\ln x = t$, so that $x = e^t$, and thus $dx = e^t dt$. This gives the following

$$I = \int e^t \sin^2 t\, dt = \frac{1}{2} \int (1 - \cos(2t))e^t dt = \frac{1}{2}\int e^t dt - \frac{1}{2}\int e^t \cos(2t)dt$$

$$= \frac{1}{2}e^t - \frac{1}{2}\left[ \frac{e^t}{5}(\cos(2t) + 2\sin(2t)) \right] + C,$$

where $C \in \mathbb{R}$.

Remember that (see (1.1))

$$\int e^{ax}\cos(bx)dx = \frac{e^{ax}}{a^2 + b^2}(a\cos(bx) + b\sin(bx)) + C.$$

But $t = \ln x$, therefore we have

$$I = \frac{1}{2}x - \frac{1}{10}x\cos(2\ln x) - \frac{1}{5}x\sin(2\ln x) + C,$$

which is the same as in (3.2)!.

$$\boxed{I_6 = \int \frac{dx}{1 + 3e^x}}$$

**Solution.**

For the given integral we can write the following

$$I_6 = \int \frac{dx}{e^x(3 + e^{-x})} = \int \frac{e^{-x}}{3 + e^{-x}}dx = -\int \frac{-e^{-x}}{3 + e^{-x}}dx$$

$$= -\ln(3 + e^{-x}) + C,$$

where $C \in \mathbb{R}$.

$$I_7 = \int_{\pi/4}^{\pi/3} \frac{dx}{\sin^3 x \cos^5 x}$$

**Solution.**

For the given integral we can write the following

$$I_7 = \int_{\pi/4}^{\pi/3} \frac{1/\cos^8 x}{\sin^3 x / \cos^3 x} dx = \int_{\pi/4}^{\pi/3} \frac{1/\cos^6 x}{\tan^3 x} \frac{dx}{\cos^2 x}$$
$$= \int_{\pi/4}^{\pi/3} \frac{\left(1/\cos^2 x\right)^3}{\tan^3 x} \frac{dx}{\cos^2 x} = \int_{\pi/4}^{\pi/3} \frac{\left(1 + \tan^2 x\right)^3}{\tan^3 x} \frac{dx}{\cos^2 x}.$$

For the last integral, let us use the substitution $\tan x = t$, so that $\frac{dx}{\cos^2 x} = dt$. If $x = \pi/4$, then $t = 1$, and if $x = \pi/3$, then $t = \sqrt{3}$. With these new values, we get the following

$$I_7 = \int_1^{\sqrt{3}} \frac{(1 + t^2)^3}{t^3} dt = \int_1^{\sqrt{3}} \frac{t^6 + 4t^4 + 3t^2 + 1}{t^3} dt$$
$$= \int_1^{\sqrt{3}} \left( t^3 + 3t + \frac{3}{t} + \frac{1}{t^3} \right) dt$$
$$= \left[ \frac{t^4}{4} + \frac{3}{2}t^2 + 3\ln|t| - \frac{1}{2t^2} \right]_1^{\sqrt{3}} = \frac{16}{3} + 3\ln(\sqrt{3}).$$

$$I_8 = \int_1^{\infty} \frac{dx}{x\sqrt{x^4 - 1}}$$

**Solution 1.**

For the given integral we can write the following

$$I_8 = \lim_{\substack{a \to 1^+ \\ b \to \infty}} \int_a^b \frac{dx}{x\sqrt{x^4 - 1}}.$$

Let us calculate the indefinite integral

$$I = \int \frac{dx}{x\sqrt{x^4 - 1}}. \tag{3.3}$$

For this integral, let us use the substitution $\sqrt{x^4 - 1} = t$, so that $x = (1 + t^2)^{\frac{1}{4}}$. Thus, $dx = \frac{1}{2}t(1 + t^2)^{-\frac{3}{4}} dt$. This gives the following

$$I = \frac{1}{2} \int \frac{t(1 + t^2)^{-3/4}}{t(1 + t^2)^{1/4}} dt = \frac{1}{2} \int \frac{dt}{1 + t^2} = \frac{1}{2} \arctan(t) + C$$

$$= \frac{1}{2} \arctan\left(\sqrt{x^4 - 1}\right) + C,$$

where $C \in \mathbb{R}$. Therefore, for the given integral $I_8$, we get

$$I_8 = \lim_{\substack{a \to 1^+ \\ b \to \infty}} \left[\frac{1}{2} \arctan\left(\sqrt{x^4 - 1}\right)\right]_a^b$$

$$= \frac{1}{2} \lim_{\substack{a \to 1^+ \\ b \to \infty}} \left(\arctan\sqrt{b^4 - 1} - \arctan\sqrt{a^4 - 1}\right) = \frac{\pi}{4}.$$

## Solution 2.

For the integral $I$ in (3.3), we have

$$I = \int \frac{dx}{x\sqrt{x^4 - 1}} = \int \frac{dx}{xx^2\sqrt{1 - x^{-4}}} = \int \frac{dx}{x^3\sqrt{1 - \frac{1}{x^4}}}.$$

For the last integral, let us use the substitution $1 - \frac{1}{x^4} = t$, so that $x^{-4} = 1 - t$. Thus,

$$-4x^{-5} dx = -dt \implies \frac{dx}{x^3 x^2} = \frac{dt}{4} \implies \frac{dx}{x^3} = \frac{dt}{4\sqrt{1 - t}}.$$

This gives the following

$$I = \int \frac{dt}{4\sqrt{1 - t}\sqrt{t}} = \frac{1}{4} \int \frac{dt}{\sqrt{t - t^2}} = \frac{1}{4} \int \frac{dt}{\sqrt{-(t^2 - t + \frac{1}{4} - \frac{1}{4})}}$$

$$= \frac{1}{4} \int \frac{dt}{\sqrt{\frac{1}{4} - (t - \frac{1}{2})^2}} = \frac{1}{4} \arcsin(2t - 1) + C = \frac{1}{4} \arcsin\left(1 - \frac{2}{x^4}\right) + C,$$

where $C \in \mathbb{R}$. Therefore, for the integral $I_8$, we find

$$I_8 = \lim_{\substack{a \to 1^+ \\ b \to \infty}} \left[\frac{1}{4} \arcsin\left(1 - \frac{2}{x^4}\right)\right]_a^b$$

$$= \frac{1}{4} \left(\lim_{b \to \infty} \arcsin\left(1 - \frac{2}{b^4}\right) - \arcsin\left(1 - \frac{2}{a^4}\right)\right)$$

$$= \frac{1}{4} (\arcsin(1) - \arcsin(-1)) = \frac{1}{4}\left(\frac{\pi}{2} + \frac{\pi}{2}\right) = \frac{\pi}{4}.$$

$$I_9 = \int \frac{dx}{x(x^5 + 1)}$$

**Solution.**

For the given integral we can write the following

$$I_9 = \int \frac{dx}{xx^5 \left(1 + \frac{1}{x^5}\right)} = \int \frac{dx}{x^6 \left(1 + \frac{1}{x^5}\right)}.$$

For the last integral, let us use the substitution $1/x^5 = t$, so that $dx/x^6 = -dt/5$. This gives the following

$$I_9 = -\frac{1}{5} \int \frac{dt}{1+t} = -\frac{1}{5} \ln|1 + t| + C = -\frac{1}{5} \ln\left|1 + \frac{1}{x^5}\right| + C$$

$$= -\frac{1}{5} \ln\left|\frac{x^5 + 1}{x^5}\right| + C = \frac{1}{5} \ln\left|\frac{x^5}{x^5 + 1}\right| + C,$$

where $C \in \mathbb{R}$.

$$I_{10} = \int_0^{\pi/4} \sqrt{\tan x}\, dx$$

**Solution.**

Let us use the substitution $\tan x = t$, so that $dx = \frac{dt}{1+t^2}$. If $x = 0$, then $t = 0$, and if $x = \pi/4$, then $t = 1$. With these new values, we get the following

$$I_{10} = \int_0^1 \frac{\sqrt{t}}{1+t^2}\, dt.$$

Now, let us use the substitution $t = u^2$, so that $dt = 2u\, du$. If $t = 0$, then $u = 0$, and if $t = 1$, then $u = 1$. This gives the following integral

$$I_{10} = \int_0^1 \frac{2u^2}{1+u^4}\, du. \tag{3.4}$$

For this integral, we have

$$u^4 + 1 = u^4 + 2u^2 + 1 - 2u^2 = (u^2 + 1)^2 - \left(\sqrt{2}u\right)^2$$

$$= \left(u^2 + \sqrt{2}u + 1\right)\left(u^2 - \sqrt{2}u + 1\right).$$

Thus,

$$\frac{2u^2}{u^4 + 1} = \frac{Au + B}{u^2 + \sqrt{2}u + 1} + \frac{Cu + D}{u^2 - \sqrt{2}u + 1}. \tag{3.5}$$

For the constants $A, B, C, D$, we have

$$2u^2 = (Au + B)\left(u^2 - \sqrt{2}u + 1\right) + (Cu + D)\left(u^2 + \sqrt{2}u + 1\right). \qquad (3.6)$$

By the corresponding between the sides of the equality (3.6), we find the following linear system of equations

$$\begin{cases} A + C = 0, \\ -\sqrt{2}A + B + \sqrt{2}C + D = 2, \\ A - \sqrt{2}B + C + \sqrt{2}D = 0, \\ B + D = 0. \end{cases}$$

By solving these equations, we find

$$A = -\frac{1}{\sqrt{2}}, \quad B = 0, \quad C = \frac{1}{\sqrt{2}}, \quad D = 0.$$

Therefore

$$\begin{aligned}
\int \frac{2u^2}{1 + u^4} du &= -\frac{1}{\sqrt{2}} \int \frac{u}{u^2 + \sqrt{2}u + 1} du + \frac{1}{\sqrt{2}} \int \frac{u}{u^2 - \sqrt{2}u + 1} du \\
&= -\frac{1}{\sqrt{2}} \cdot \frac{1}{2} \int \frac{2u + \sqrt{2} - \sqrt{2}}{u^2 + \sqrt{2}u + 1} du + \frac{1}{\sqrt{2}} \cdot \frac{1}{2} \int \frac{2u - \sqrt{2} + \sqrt{2}}{u^2 - \sqrt{2}u + 1} du \\
&= -\frac{1}{2\sqrt{2}} \int \frac{2u + \sqrt{2}}{u^2 + \sqrt{2}u + 1} du + \frac{1}{2} \int \frac{du}{u^2 + \sqrt{2}u + 1} \\
&\quad + \frac{1}{2\sqrt{2}} \int \frac{2u - \sqrt{2}}{u^2 - \sqrt{2}u + 1} du + \frac{1}{2} \int \frac{du}{u^2 - \sqrt{2}u + 1} \\
&= \frac{1}{2\sqrt{2}} \left( \int \frac{2u - \sqrt{2}}{u^2 - \sqrt{2}u + 1} du - \int \frac{2u + \sqrt{2}}{u^2 + \sqrt{2}u + 1} du \right) \\
&\quad + \frac{1}{2} \left( \int \frac{du}{\left(u + \frac{1}{\sqrt{2}}\right)^2 + \left(\frac{1}{\sqrt{2}}\right)^2} + \int \frac{du}{\left(u - \frac{1}{\sqrt{2}}\right)^2 + \left(\frac{1}{\sqrt{2}}\right)^2} \right) \\
&= \frac{1}{2\sqrt{2}} \ln \left| \frac{u^2 - \sqrt{2}u + 1}{u^2 + \sqrt{2}u + 1} \right| + \frac{1}{\sqrt{2}} \arctan\left(\sqrt{2}u + 1\right) \\
&\quad + \frac{1}{\sqrt{2}} \arctan\left(\sqrt{2}u - 1\right) + C',
\end{aligned}$$

where $C' \in \mathbb{R}$. Therefore, for the integral $I_{10}$ in (3.4), we have

$$I_{10} = \left[ \frac{1}{2\sqrt{2}} \ln \left| \frac{u^2 - \sqrt{2}u + 1}{u^2 + \sqrt{2}u + 1} \right| + \frac{1}{\sqrt{2}} \arctan\left(\sqrt{2}u + 1\right) + \right.$$

$$\left. + \frac{1}{\sqrt{2}} \arctan\left(\sqrt{2}u - 1\right) \right]_0^1$$

$$= \frac{1}{2\sqrt{2}} \ln\left(\frac{2 - \sqrt{2}}{2 + \sqrt{2}}\right) + \frac{\pi}{2\sqrt{2}}.$$

Remember that

$$\arctan(a) + \arctan(b) = \arctan\left(\frac{a + b}{1 - ab}\right).$$

**Remark.** For the integral

$$J := \int \frac{2u^2}{1 + u^4} du,$$

we can write the following

$$J = \int \frac{2u^2}{1 + u^4} du = \int \frac{2}{u^2 + \frac{1}{u^2}} du = \int \frac{1 + \frac{1}{u^2} + 1 - \frac{1}{u^2}}{u^2 + \frac{1}{u^2}} du$$

$$= \underbrace{\int \frac{1 + \frac{1}{u^2}}{\left(u - \frac{1}{u}\right)^2 + 2} du}_{:=J_1} + \underbrace{\int \frac{1 - \frac{1}{u^2}}{\left(u + \frac{1}{u}\right)^2 - 2} du}_{:=J_2}.$$

For the integral $J_1$, let us use the substitution

$$u - \frac{1}{u} = y \implies \left(1 + \frac{1}{u^2}\right) du = dy,$$

and for the integral $J_2$, let we use the substitution

$$u + \frac{1}{u} = z \implies \left(1 - \frac{1}{u^2}\right) du = dz.$$

From These substitutions we get

$$J = \int \frac{2u^2}{1 + u^4} du = \int \frac{dy}{(\sqrt{2})^2 + y^2} - \int \frac{dz}{(\sqrt{2})^2 - z^2}$$

$$= \frac{1}{\sqrt{2}} \arctan(y) - \frac{1}{2\sqrt{2}} \ln\left|\frac{\sqrt{2} + z}{\sqrt{2} - z}\right| + C$$

$$= \frac{1}{\sqrt{2}} \arctan\left(u - \frac{1}{u}\right) - \frac{1}{2\sqrt{2}} \ln\left|\frac{\sqrt{2} + u + \frac{1}{u}}{\sqrt{2} - u - \frac{1}{u}}\right| + C$$

$$= \frac{1}{\sqrt{2}} \arctan\left(u - \frac{1}{u}\right) - \frac{1}{2\sqrt{2}} \ln\left|\frac{\sqrt{2}u + u^2 + 1}{\sqrt{2}u - u^2 - 1}\right| + C,$$

where $C \in \mathbb{R}$. Therefore, for the integral $I_{10}$ in (3.4), we have

$$I_{10} = \left[ \frac{1}{\sqrt{2}} \arctan\left(u - \frac{1}{u}\right) - \frac{1}{2\sqrt{2}} \ln\left|\frac{\sqrt{2}u + u^2 + 1}{\sqrt{2}u - u^2 - 1}\right| \right]_{u \to 0}^{1}$$

$$= -\frac{1}{2\sqrt{2}} \ln\left|\frac{2 + \sqrt{2}}{2 - \sqrt{2}}\right|$$

$$- \lim_{u \to 0} \left(\frac{1}{\sqrt{2}} \arctan\left(u - \frac{1}{u}\right) - \frac{1}{2\sqrt{2}} \ln\left|\frac{\sqrt{2}u + u^2 + 1}{\sqrt{2}u - u^2 - 1}\right|\right)$$

$$= \frac{1}{2\sqrt{2}} \ln\left(\frac{2 - \sqrt{2}}{2 + \sqrt{2}}\right) + \frac{\pi}{2\sqrt{2}}.$$

$$\boxed{I_{11} = \int_0^1 \frac{\ln(1 + x)}{1 + x^2} dx}$$

**Solution.**

Let us use the substitution $x = \tan t$, so that $dx = dt/\cos^2 t$. If $x = 0$, then $t = 0$, and if $x = 1$, then $t = \pi/4$. With these new values, we get the following

$$I_{11} = \int_0^{\pi/4} \frac{\ln(1 + \tan t)}{1 + \tan^2 t} \frac{dt}{\cos^2 t} = \int_0^{\pi/4} \ln(1 + \tan t) dt. \qquad (3.7)$$

For the last integral, let us use the substitution $y = \frac{\pi}{4} - t$, so that $dy = -dt$. If $t = 0$, then $y = \pi/4$, and if $t = \pi/4$, then $y = 0$. With these new values, we get the following

$$I_{11} = -\int_{\pi/4}^0 \ln\left(1 + \tan\left(\frac{\pi}{4} - y\right)\right) dy = \int_0^{\pi/4} \ln\left(1 + \tan\left(\frac{\pi}{4} - y\right)\right) dy.$$

But

$$\tan\left(\frac{\pi}{4} - y\right) = \frac{\tan(\pi/4) - \tan y}{1 + \tan(\pi/4)\tan y} = \frac{1 - \tan y}{1 + \tan y}.$$

Thus, we have

$$I_{11} = \int_0^{\pi/4} \ln\left(1 + \frac{1 - \tan y}{1 + \tan y}\right) dy = \int_0^{\pi/4} \ln\left(\frac{2}{1 + \tan y}\right) dy$$

$$= \int_0^{\pi/4} \ln\left(\frac{2}{1 + \tan t}\right) dt. \qquad (3.8)$$

Now, by adding (3.7) and (3.8), we find

$$2I_{11} = \int_0^{\pi/4} \ln\left(1 + \tan t\right) dt + \int_0^{\pi/4} \ln\left(\frac{2}{1 + \tan t}\right) dt$$

$$= \int_0^{\pi/4} \ln\left((1 + \tan t)\left(\frac{2}{1 + \tan t}\right)\right) dt$$

$$= \ln 2 \int_0^{\pi/4} dt = \frac{\pi}{4} \ln 2.$$

Therefore

$$I_{11} = \frac{\pi}{8} \ln 2.$$

$$I_{12} = \int_{64}^{729} \frac{\sqrt{x}}{\sqrt{x} - \sqrt[3]{x}} dx$$

**Solution.**

Let us use the substitution $x = t^6$, so that $dx = 6t^5 dt$. If $x = 64$, then $t = 2$, and if $x = 729$, then $t = 3$. With these new values, we get the following

$$I_{12} = 6 \int_2^3 \frac{t^8}{t^3 - t^2} dt = 6 \int_2^3 \frac{t^6}{t - 1} dt = 6 \int_2^3 \frac{((t^3)^2 - 1) + 1}{t - 1} dt$$

$$= 6 \int_2^3 \left(\frac{(t^3 - 1)(t^3 + 1)}{t - 1} + \frac{1}{t - 1}\right) dt$$

$$= 6 \int_2^3 \left(\frac{(t - 1)(t^2 + t + 1)(t^3 + 1)}{t - 1} + \frac{1}{t - 1}\right) dt$$

$$= 6 \int_2^3 \left((t^2 + t + 1)(t^3 + 1) + \frac{1}{t - 1}\right) dt$$

$$= 6 \int_2^3 \left(t^5 + t^4 + t^3 + t^2 + t + 1 + \frac{1}{t - 1}\right) dt$$

$$= 6 \left[\frac{t^6}{6} + \frac{t^5}{5} + \frac{t^4}{4} + \frac{t^3}{3} + \frac{t^2}{2} + t + \ln|t - 1|\right]_2^3 = \frac{10747}{10} + 6 \ln 2.$$

$$I_{13} = \int x^x (1 + \ln x) dx \qquad\qquad (3.9)$$

**Solution.**

For the given integral we can write the following

$$I_{13} = \int e^{x \ln x} (1 + \ln x) dx.$$

Now, let us use the substitution $x \ln x = t$, so that $(1 + \ln x)dx = dt$. This gives the following

$$I_{13} = \int e^t dt = e^t + C = e^{x \ln x} + C = e^{\ln(x^x)} + C = x^x + C,$$

where $C \in \mathbb{R}$.

$$I_{14} = \int_0^1 x^{13/2} \sqrt{1 + x^{5/2}} dx$$

### Solution.

Let us use the substitution $1 + x^{5/2} = t^2$, so that $x = (t^2 - 1)^{2/5}$. Thus, $dx = \frac{4}{5} t(t^2 - 1)^{-3/5} dt$. If $x = 0$, then $t = 1$, and if $x = 1$, then $t = \sqrt{2}$. With these new values, we get the following

$$I_{14} = \int_1^{\sqrt{2}} (t^2 - 1)^{13/5} \frac{4}{5} t(t^2 - 1)^{-3/5} dt = \frac{4}{5} \int_1^{\sqrt{2}} t^2 (t^2 - 1)^2 dt$$

$$= \frac{4}{5} \int_1^{\sqrt{2}} (t^6 - 2t^4 + t^2) dt = \frac{4}{5} \left[ \frac{t^7}{7} - \frac{2}{5} t^5 + \frac{t^3}{3} \right]_1^{\sqrt{2}} = \frac{88\sqrt{2} - 32}{525}.$$

$$I_{15} = \int_0^\infty \frac{dx}{(1 + x^2)^2}$$

### Solution.

For the given integral we can write the following

$$I_{15} = \lim_{b \to \infty} \int_0^b \frac{dx}{(1 + x^2)^2}.$$

Now, let us calculate the indefinite integral

$$I = \int \frac{dx}{(1 + x^2)^2}.$$

For this integral, we can write

$$I = \int \frac{1 + x^2 - x^2}{(1 + x^2)^2} dx = \int \frac{dx}{1 + x^2} - \underbrace{\int x \frac{x}{(x^2 + 1)^2} dx}_{:=J}.$$

Now, for the integral $J$, by using the integration by parts, let us assume

$$u = x \implies du = dx, \quad \frac{x}{(x^2 + 1)^2} dx = dv \implies v = -\frac{1}{2(1 + x^2)}.$$

Thus, we have

$$I = \int \frac{dx}{1+x^2} + \frac{x}{2(1+x^2)} - \frac{1}{2} \int \frac{dx}{1+x^2} = \frac{1}{2} \arctan x + \frac{x}{2(1+x^2)} + C,$$

where $C \in \mathbb{R}$. Therefore, for the given integral $I_{15}$, we get

$$I_{15} = \lim_{b \to \infty} \left[ \frac{1}{2} \arctan x + \frac{x}{2(1+x^2)} \right]_0^b = \lim_{b \to \infty} \left( \frac{1}{2} \arctan b + \frac{b}{2(1+b^2)} \right) = \frac{\pi}{4}.$$

$$\boxed{I_{16} = \int_0^1 \frac{dx}{x^4 - 13x^2 + 36}}$$

**Solution.**

According to the method of partial fractions, we have

$$f(x) := \frac{1}{x^4 - 13x^2 + 36} = \frac{1}{(x^2 - 9)(x^2 - 4)} = \frac{1}{(x-3)(x+3)(x-2)(x+2)}$$
$$= \frac{A}{x-3} + \frac{B}{x+3} + \frac{C}{x-2} + \frac{D}{x+2}.$$

Where

$$A = \frac{1}{(x+3)(x-2)(x+2)} \bigg|_{x=3} = \frac{1}{30},$$

$$B = \frac{1}{(x-3)(x-2)(x+2)} \bigg|_{x=-3} = -\frac{1}{30},$$

$$C = \frac{1}{(x-3)(x+3)(x+2)} \bigg|_{x=2} = -\frac{1}{20},$$

$$D = \frac{1}{(x-3)(x+3)(x-2)} \bigg|_{x=-2} = \frac{1}{20}.$$

Thus,

$$f(x) = \frac{1}{30} \left( \frac{1}{x-3} - \frac{1}{x+3} \right) + \frac{1}{20} \left( \frac{1}{x+2} - \frac{1}{x-2} \right).$$

Therefore, for the given integral $I_{16}$, we have

$$I_{16} = \left[ \frac{1}{30} \left( \ln|x-3| - \ln|x+3| \right) + \frac{1}{20} \left( \ln|x+2| - \ln|x-2| \right) \right]_0^1$$
$$= \left[ \frac{1}{30} \ln \left| \frac{x-3}{x+3} \right| + \frac{1}{20} \ln \left| \frac{x+2}{x-2} \right| \right]_0^1 = \frac{1}{20} \ln 3 - \frac{1}{30} \ln 2.$$

$$I_{17} = \int \frac{\ln(\ln x)}{x} dx$$

**Solution.**

Let us use the substitution $\ln x = t$, so that $dx/x = dt$. This gives the following

$$I_{17} = \int \ln t\, dt.$$

By using the integration by parts, let us assume

$$u = \ln t \Longrightarrow du = \frac{dt}{t}, \quad dv = dt \Longrightarrow v = t.$$

Thus, we have

$$I_{17} = t \ln t - \int dt = t \ln t - t + C = \ln x\,(\ln(\ln x)) - \ln x + C,$$

where $C \in \mathbb{R}$.

$$I_{18} = \int \frac{1 + \cot x}{1 - \cot x} dx$$

**Solution.**

For the given integral we can write the following

$$I_{18} = \int \frac{1 + \frac{\cos x}{\sin x}}{1 - \frac{\cos x}{\sin x}} dx = \int \frac{\sin x + \cos x}{\sin x - \cos x} dx = \ln|\sin x - \cos x| + C,$$

where $C \in \mathbb{R}$.

$$I_{19} = \int \frac{\cos x + x \sin x}{x(x + \cos x)} dx$$

**Solution 1.**

For the given integral we can write the following

$$I_{19} = \int \frac{x + \cos x + x \sin x - x}{x(x + \cos x)} dx$$

$$= \int \frac{x + \cos x}{x(x + \cos x)} dx + \int \frac{x \sin x - x}{x(x + \cos x)} dx$$

$$= \int \frac{dx}{x} - \int \frac{x(1 - \sin x)}{x(x + \cos x)} dx = \ln|x| - \ln|x + \cos x| + C,$$

where $C \in \mathbb{R}$.

**Solution 2.**

For the given integral we can write the following

$$I_{19} = \int \frac{\cos x + x \sin x}{x^2 \left(1 + \frac{\cos x}{x}\right)} dx.$$

Now, let us use the substitution

$$\frac{\cos x}{x} = t \implies \frac{-x \sin x - \cos x}{x^2} dx = dt \implies \frac{x \sin x + \cos x}{x^2} dx = -dt.$$

This gives the following

$$I_{19} = -\int \frac{dt}{1+t} = -\ln|1+t| + C = -\ln\left|1 + \frac{\cos x}{x}\right| + C$$

$$= -\ln\left|\frac{x + \cos x}{x}\right| + C = \ln|x| - \ln|x + \cos x| + C,$$

where $C \in \mathbb{R}$.

$$I_{20} = \int_0^{\pi/2} \frac{dx}{\sin x + \sec x}$$

**Solution 1.**

For the given integral we can write the following

$$I_{20} = \int_0^{\pi/2} \frac{\cos x}{1 + \sin x \cos x} dx. \qquad (3.10)$$

By using the King property of integration (1.13), we find

$$I_{20} = \int_0^{\pi/2} \frac{\cos\left(\frac{\pi}{2} - x\right)}{1 + \sin\left(\frac{\pi}{2} - x\right)\cos\left(\frac{\pi}{2} - x\right)} dx = \int_0^{\pi/2} \frac{\sin x}{1 + \sin x \cos x} dx. \qquad (3.11)$$

By adding (3.10) and (3.11), we find

$$2I_{20} = \int_0^{\pi/2} \frac{\sin x + \cos x}{1 + \sin x \cos x} dx = 2 \int_0^{\pi/2} \frac{\sin x + \cos x}{2 + 2 \sin x \cos x} dx$$

$$= 2 \int_0^{\pi/2} \frac{\sin x + \cos x}{3 - (\cos x - \sin x)^2} dx.$$

Now, for the last integral, let us use the substitution $\sin x - \cos x = t$, so that $(\sin x + \cos x) dx = dt$. If $x = 0$, then $t = -1$, and if $x = \pi/2$, then $t = 1$. With these new values, we get the following

$$2I_{20} = 2\int_{-1}^{1}\frac{dt}{3-t^2} = 2\frac{1}{2\sqrt{3}}\left[\ln\left|\frac{\sqrt{3}+t}{\sqrt{3}-t}\right|\right]_{-1}^{1}$$

$$= \frac{1}{\sqrt{3}}\left(\ln\left(\sqrt{3}+1\right) - \ln\left(\sqrt{3}-1\right) - \ln\left(\sqrt{3}-1\right) + \ln\left(\sqrt{3}+1\right)\right)$$

$$= \frac{2}{\sqrt{3}}\left(\ln\left(\sqrt{3}+1\right) - \ln\left(\sqrt{3}-1\right)\right) = \frac{2}{\sqrt{3}}\ln\left(2+\sqrt{3}\right).$$

Therefore

$$I_{20} = \frac{1}{\sqrt{3}}\ln\left(2+\sqrt{3}\right).$$

**Solution 2.**

For the given integral we can write the following

$$I_{20} = \int_{0}^{\pi/2}\frac{\cos x}{1+\sin x\cos x}\,dx = \int_{0}^{\pi/2}\frac{2\cos x}{2+\sin(2x)}\,dx.$$

Now, let us calculate the indefinite integral

$$J := \int\frac{2\cos x}{2+\sin(2x)}\,dx.$$

For this, we have

$$J = \int\frac{\cos x + \sin x + \cos x - \sin x}{2+\sin(2x)}\,dx$$

$$= \int\frac{\cos x + \sin x}{2+\sin(2x)}\,dx + \int\frac{\cos x - \sin x}{2+\sin(2x)}\,dx$$

$$= \underbrace{\int\frac{\cos x + \sin x}{3-(\sin x-\cos x)^2}\,dx}_{:=J_1} + \underbrace{\int\frac{\cos x - \sin x}{1+(\sin x+\cos x)^2}\,dx}_{:=J_2}.$$

For $J_1$, let us use the substitution $\sin x - \cos x = t$, thus $(\cos x + \sin x)dx = dt$, and For $J_2$, let us use the substitution $\sin x + \cos x = u$, thus $(\cos x - \sin x)dx = du$. This gives the following

$$J = \int\frac{dt}{3-t^2} + \int\frac{du}{1+u^2} = \frac{1}{2\sqrt{3}}\ln\left|\frac{\sqrt{3}+t}{\sqrt{3}-t}\right| + \arctan u + C$$

$$= \frac{1}{2\sqrt{3}}\ln\left|\frac{\sqrt{3}+\sin x-\cos x}{\sqrt{3}-\sin x+\cos x}\right| + \arctan(\sin x+\cos x) + C,$$

where $C \in \mathbb{R}$. Therefore, for the given integral $I_{20}$, we get

$$I_{20} = \left[ \frac{1}{2\sqrt{3}} \ln \left| \frac{\sqrt{3} + \sin x - \cos x}{\sqrt{3} - \sin x + \cos x} \right| + \arctan(\sin x + \cos x) \right]_0^{\pi/2}$$

$$= \frac{1}{2\sqrt{3}} \ln \left( \frac{\sqrt{3}+1}{\sqrt{3}-1} \right) + \frac{\pi}{4} - \frac{1}{2\sqrt{3}} \ln \left( \frac{\sqrt{3}-1}{\sqrt{3}+1} \right) - \frac{\pi}{4}$$

$$= \frac{1}{2\sqrt{3}} \ln \left( \frac{\sqrt{3}+1}{\sqrt{3}-1} \frac{\sqrt{3}+1}{\sqrt{3}-1} \right) = \frac{1}{2\sqrt{3}} \ln \left( \frac{\sqrt{3}+1}{\sqrt{3}-1} \right)^2$$

$$= \frac{1}{\sqrt{3}} \ln \left( \frac{\sqrt{3}+1}{\sqrt{3}-1} \right) = \frac{1}{\sqrt{3}} \ln \left( 2 + \sqrt{3} \right).$$

$$\boxed{I_{21} = \int_0^\infty \frac{dx}{\sqrt{1 + e^x + e^{2x}}}}$$

**Solution.**

For the given integral we can write the following

$$I_{21} = \int_0^\infty \frac{dx}{e^x \sqrt{1 + e^{-x} + (e^{-x})^2}} = \int_0^\infty \frac{e^{-x}}{\sqrt{1 + e^{-x} + (e^{-x})^2}} dx.$$

For the last integral, let us use the substitution $e^{-x} = t$, so that $e^{-x} dx = -dt$. If $x = 0$, then $t = 1$, and if $x \to \infty$, then $t \to 0$. With these new values, we get the following

$$I_{21} = -\int_1^0 \frac{dt}{\sqrt{t^2 + t + 1}} = \int_0^1 \frac{dt}{\sqrt{t^2 + t + \frac{1}{4} + \frac{3}{4}}} = \int_0^1 \frac{dt}{\sqrt{\left( t + \frac{1}{2} \right)^2 + \frac{3}{4}}}$$

$$= \left[ \ln \left| t + \frac{1}{2} + \sqrt{t^2 + t + 1} \right| \right]_0^1 = \ln \left( \frac{3 + 2\sqrt{3}}{2} \right) - \ln \left( \frac{3}{2} \right)$$

$$= \ln(3 + 2\sqrt{3}) - \ln 3.$$

$$\boxed{I_{22} = \int_0^1 x^3 e^{x^2} dx}$$

**Solution.**

For the given integral we can write the following

$$I_{22} = \int_0^1 x x^2 e^{x^2} dx.$$

Let us use the substitution $x^2 = t$, so that $2x dx = dt$. If $x = 0$, then $t = 0$, and if $x = 1$, then $t = 1$. With these new values, we get the following

$$I_{22} = \frac{1}{2} \int_0^1 te^t dt = \frac{1}{2} \left( [te^t]_0^1 - \int_0^1 e^t dt \right) = \frac{1}{2} (e - (e-1)) = \frac{1}{2}.$$

Where we used the integration by parts, by assuming $u = t$ and $e^t dt = dv$.

$$I_{23} = \int_0^1 \sqrt{1 + x\sqrt{1 + x\sqrt{1 + x\sqrt{\dots}}}}\, dx$$

**Solution.**

Let we set

$$t = \sqrt{1 + x\sqrt{1 + x\sqrt{1 + x\sqrt{\dots}}}} > 0, \quad \forall x \in [0, 1].$$

Then

$$t^2 = 1 + x\sqrt{1 + x\sqrt{1 + x\sqrt{1 + x\sqrt{\dots}}}} = 1 + xt,$$

i.e.

$$t^2 - xt - 1 = 0.$$

By solving this quadratic equation, for $t$, we find the following solutions:

$$t_1 = \frac{x}{2} - \frac{1}{2}\sqrt{x^2 + 4} < 0, \quad t_2 = \frac{x}{2} + \frac{1}{2}\sqrt{x^2 + 4} > 0, \quad \forall x \in [0, 1].$$

Therefore, we have the following integral (for $t > 0$)

$$I_{23} = \int_0^1 \left( \frac{x}{2} + \frac{1}{2}\sqrt{x^2 + 4} \right) dx. \tag{3.12}$$

Now, let us calculate the following indefinite integral

$$I = \int \sqrt{x^2 + 4}\, dx.$$

For this, we have

$$I = 2 \int \sqrt{1 + (x/2)^2}\, dx.$$

Now, let us use the substitution $x/2 = \sinh t$, so that $dx = 2\cosh t dt$, and $t = \operatorname{arcsinh}(x/2)$. This gives the following

$$I = 2 \int \sqrt{1 + \sinh^2 t} \, (2 \cosh t) dt = 4 \int \cosh^2 t \, dt = 4 \int \frac{1 + \cosh(2t)}{2} dt$$

$$= 2 \left( t + \frac{1}{2} \sinh(2t) \right) + C = 2t + \sinh(2t) + C$$

$$= 2 \operatorname{arcsinh} \left( \frac{x}{2} \right) + \sinh \left( 2 \operatorname{arcsinh} \left( \frac{x}{2} \right) \right) + C$$

$$= 2 \operatorname{arcsinh} \left( \frac{x}{2} \right) + 2 \sinh \left( \operatorname{arcsinh} \left( \frac{x}{2} \right) \right) \cosh \left( \operatorname{arcsinh} \left( \frac{x}{2} \right) \right) + C$$

$$= 2 \operatorname{arcsinh} \left( \frac{x}{2} \right) + x \sqrt{1 + \left( \frac{x}{2} \right)^2} + C = 2 \operatorname{arcsinh} \left( \frac{x}{2} \right) + \frac{x}{2} \sqrt{x^2 + 4} + C,$$

where $C \in \mathbb{R}$.

As a result of the previous, from (3.12), we have

$$I_{23} = \int_0^1 \left( \frac{x}{2} + \frac{1}{2} \sqrt{x^2 + 4} \right) dx = \left[ \frac{x^2}{4} + \operatorname{arcsinh} \left( \frac{x}{2} \right) + \frac{x}{4} \sqrt{x^2 + 4} \right]_0^1$$

$$= \frac{1 + \sqrt{5}}{4} + \operatorname{arcsinh} \left( \frac{1}{2} \right) = \frac{1 + \sqrt{5}}{4} + \ln \left( \frac{1}{2} + \sqrt{1 + \frac{1}{4}} \right)$$

$$= \frac{1 + \sqrt{5}}{4} + \ln \left( \frac{1 + \sqrt{5}}{2} \right).$$

Where we used the fact

$$\operatorname{arcsinh}(x) = \ln \left( x + \sqrt{1 + x^2} \right), \quad \forall x \in \mathbb{R}.$$

$$\boxed{I_{24} = \int \left( \frac{1}{\ln x} - \frac{1}{\ln^2 x} \right) dx}$$

**Solution 1.**

We note that

$$\left( \frac{x}{\ln x} \right)' = \frac{\ln x - 1}{\ln^2 x} = \frac{1}{\ln x} - \frac{1}{\ln^2 x}, \quad \forall x \in ]0, \infty[ \backslash \{1\}.$$

Therefore, for the given integral, we get

$$I_{24} = \int \left( \frac{x}{\ln x} \right)' dx = \frac{x}{\ln x} + C,$$

where $C \in \mathbb{R}$.

**Solution 2.**

Let us use the substitution $\ln x = t$, so that $x = e^t$ and $dx = e^t dt$. This gives the following

$$I_{24} = \int \left( \frac{1}{t} - \frac{1}{t^2} \right) e^t dt = \underbrace{\int \frac{e^t}{t} dt}_{:=J} - \int \frac{e^t}{t^2} dt.$$

For the integral $J$, by using the integration by parts, we assume

$$u = \frac{1}{t} \Longrightarrow du = -\frac{dt}{t^2}, \quad dv = e^t dt \Longrightarrow v = e^t.$$

Thus,

$$I_{24} = \underbrace{\frac{e^t}{t} + \int \frac{e^t}{t^2} dt}_{J} - \int \frac{e^t}{t^2} dt = \frac{e^t}{t} + C = \frac{e^{\ln x}}{\ln x} + C = \frac{x}{\ln x} + C,$$

where $C \in \mathbb{R}$.

$$\boxed{I_{25} = \int_1^2 (x-1)^{1/2}(2-x)^{1/2} dx}$$

**Solution.**

For the given integral we can write the following

$$I_{25} = \int_1^2 \sqrt{(x-1)(2-x)} dx = \int_1^2 \sqrt{-x^2 + 3x - 2} dx$$

$$= \int_1^2 \sqrt{-\left(x^2 - 3x + \frac{9}{4} - \frac{9}{4}\right) - 2} dx = \int_1^2 \sqrt{\frac{1}{4} - \left(x - \frac{3}{2}\right)^2} dx$$

$$= \frac{1}{2} \int_1^2 \sqrt{1 - (2x-3)^2} dx.$$

For the last integral, let us use the substitution $2x - 3 = t$, so that $2dx = dt$. If $x = 1$, then $t = -1$, and if $x = 2$, then $t = 1$. With these new values, we get the following

$$I_{25} = \frac{1}{4} \int_{-1}^1 \sqrt{1 - t^2} dt = \frac{1}{2} \int_0^1 \sqrt{1 - t^2} dt,$$

where the function $t \mapsto \sqrt{1 - t^2}$ is an even.

Now, let us use the substitution $t = \sin y$, so that $dt = \cos y dy$. If $t = 0$, then $y = 0$, and if $t = 1$, then $y = \pi/2$. With these new values, we get the following

$$I_{25} = \frac{1}{2} \int_0^{\pi/2} \sqrt{1 - \sin^2 y} \cos y \, dy = \frac{1}{2} \int_0^{\pi/2} \cos^2 y \, dy$$

$$= \frac{1}{4} \int_0^{\pi/2} (1 + \cos(2y)) dy = \frac{1}{4} \left[ y + \frac{1}{2} \sin(2y) \right]_0^{\pi/2}$$

$$= \frac{1}{4} \left( \frac{\pi}{2} + \frac{1}{2} \sin \pi - 0 \right) = \frac{\pi}{8}.$$

# Chapter 4

# The Solutions to the 2011 MIT Integration Bee, Qualifying Test

$$I_1 = \int \frac{x^6 - 1}{x^4 + x^3 - x - 1} dx$$

**Solution.**
For the given integral we can write the following

$$I_1 = \int \frac{(x^3)^2 - 1}{x^3(x+1) - (x+1)} dx = \int \frac{(x^3 - 1)(x^3 + 1)}{(x+1)(x^3 - 1)} dx$$

$$= \int \frac{(x+1)(x^2 - x + 1)}{x+1} dx = \int (x^2 - x + 1) dx$$

$$= \frac{x^3}{3} - \frac{x^2}{2} + x + C,$$

where $C \in \mathbb{R}$.

**Remark.** For the integral $I_1$, we can divide $x^6 - 1$ by $x^4 + x^3 - x - 1$. The result of the division is $x^2 - x + 1$ and the remainder is 0, i.e. $\frac{x^6 - 1}{x^4 + x^3 - x - 1} = x^2 - x + 1$. Then we can simply calculate the given integral $I_1$.

$$I_2 = \int \left( 2 \ln x + (\ln x)^2 \right) dx$$

**Solution 1.**
Note that

$$\left( x \ln^2 x \right)' = \ln^2 x + 2 \ln x.$$

69

Thus, we have

$$I_2 = \int \left(2\ln x + \ln^2 x\right) dx = \int \left(x\ln^2 x\right)' dx = x\ln^2 x + C,$$

where $C \in \mathbb{R}$.

**Solution 2.**

For the given integral we can write the following

$$I_2 = 2\int \ln x\,dx + \underbrace{\int \ln^2 x\,dx}_{:=J}.$$

For the integral $J$, by using the integration by parts, let us assume

$$u = \ln^2 x \Longrightarrow du = \frac{2\ln x}{x}dx, \quad dv = dx \Longrightarrow v = x.$$

Thus, we have

$$J = x\ln^2 x - 2\int \frac{\ln x}{x}x\,dx = x\ln^2 x - 2\int \ln x\,dx.$$

Therefore, for the given integral, we get

$$I_2 = 2\int \ln x\,dx + x\ln^2 x - 2\int \ln x\,dx + C = x\ln^2 x + C,$$

where $C \in \mathbb{R}$.

$$\boxed{I_3 = \int \frac{2x}{\sqrt{1-x^4}}dx}$$

**Solution.**

Let us use the substitution $x^2 = t$, so that $2x\,dx = dt$. This gives the following

$$I_3 = \int \frac{dt}{\sqrt{1-t^2}} = \arcsin t + C = \arcsin(x^2) + C,$$

where $C \in \mathbb{R}$.

$$\boxed{I_4 = \int \frac{x^2+1}{x+1}dx}$$

**Solution.**

For the given integral we can write the following

$$I_4 = \int \left( x - 1 + \frac{2}{x+1} \right) dx = \frac{x^2}{2} - x + 2 \ln |x + 1| + C,$$

where $C \in \mathbb{R}$.

$$I_5 = \int \frac{\sin^3 x + \sin^2 x - 2 \sin x - 2}{\sin^2 x + 2 \sin x + 1} dx$$

**Solution.**

For the given integral we can write the following

$$I_5 = \int \frac{\sin^2 x(1 + \sin x) - 2(1 + \sin x)}{(1 + \sin x)^2} dx = \int \frac{\sin^2 x - 2}{1 + \sin x} dx$$

$$= \int \frac{(\sin^2 x - 1) - 1}{1 + \sin x} dx = \int \left( \frac{(\sin x - 1)(1 + \sin x)}{1 + \sin x} - \frac{1}{1 + \sin x} \right) dx$$

$$= \int (\sin x - 1) dx - \underbrace{\int \frac{1}{1 + \sin x}}_{:=J} .$$

For the integral $J$, let us use the substitution $\tan \left( \frac{x}{2} \right) = t$, so that $dx = \frac{2dt}{1+t^2}$. This gives the following

$$J = 2 \int \frac{dt}{t^2 + 2t + 1} = 2 \int \frac{dt}{(1+t)^2} = -\frac{2}{1+t} + C = -\frac{2}{1 + \tan(x/2)} + C,$$

where $C \in \mathbb{R}$. Therefore, we have

$$I_5 = \cos x - x + \frac{2}{1 + \tan(x/2)} + C.$$

$$I_6 = \int \sinh^{-2}(x) dx$$

**Solution.**

For the given integral we can write the following

$$I_6 = \int \frac{dx}{\sinh^2 x} = \coth x + C,$$

where $C \in \mathbb{R}$.

$$I_7 = \int \sec^4 x \tan^2 x \, dx$$

**Solution.**

For the given integral we can write the following

$$I_7 = \int \frac{1}{\cos^4 x} \frac{\sin^2 x}{\cos^2 x} dx = \int \frac{\sin^2 x}{\cos^6 x} dx = \int \frac{\sin^2 x}{\cos^2 x} \frac{1}{\cos^2 x} \frac{dx}{\cos^2 x}$$
$$= \int \tan^2 x \left(1 + \tan^2 x\right) \frac{dx}{\cos^2 x}.$$

For the last integral, let us use the substitution $\tan x = t$, so that $dx/\cos^2 x = dt$. This gives the following

$$I_7 = \int \left(t^4 + t^2\right) dt = \frac{1}{5}t^5 + \frac{1}{3}t^3 + C = \frac{1}{5}\tan^5 x + \frac{1}{3}\tan^3 x + C,$$

where $C \in \mathbb{R}$.

$$\boxed{I_8 = \int \sqrt{\csc x - \sin x}\, dx}$$

**Solution.**

For the given integral we can write the following

$$I_8 = \int \sqrt{\frac{1}{\sin x} - \sin x}\, dx = \int \sqrt{\frac{1 - \sin^2 x}{\sin x}} dx = \int \frac{\cos x}{\sqrt{\sin x}} dx$$
$$= \int \sin^{-1/2} x \cos x\, dx = 2\sqrt{\sin x} + C,$$

where $C \in \mathbb{R}$.

In our solution, we assumed that $\cos x > 0$ (Be attentive when the integral is definite!).

$$\boxed{I_9 = \int \cos^6 x\, dx}$$

**Solution.**

For the given integral we can write the following

$$I_9 = \int \left(\cos^2 x\right)^3 dx = \int \left(\frac{1 + \cos(2x)}{2}\right)^3 dx$$

$$= \frac{1}{8} \int \left(1 + \cos(2x) + 3\cos^2(2x) + \cos^3(2x)\right) dx$$

$$= \frac{1}{8} \int \left(1 + \cos(2x) + \frac{3}{2}\left(1 + \cos(4x)\right) + \cos(2x)\left(1 - \sin^2(2x)\right)\right) dx$$

$$= \frac{1}{8} \int \left(1 + \cos(2x) + \frac{3}{2} + \frac{3}{2}\cos(4x) + \cos(2x) - \cos(2x)\sin^2(2x)\right) dx$$

$$= \frac{1}{8} \int \left(\frac{5}{2} + 2\cos(2x) + \frac{3}{2}\cos(4x) - \frac{1}{2}2\cos(2x)\sin^2(2x)\right) dx$$

$$= \frac{1}{8} \left(\frac{5}{2}x + \sin(2x) + \frac{3}{8}\sin(4x) - \frac{1}{2}\frac{\sin^3(2x)}{3}\right) + C$$

$$= \frac{5}{16}x + \frac{1}{8}\sin(2x) + \frac{3}{64}\sin(4x) - \frac{1}{48}\sin^3(2x) + C,$$

where $C \in \mathbb{R}$.

**Remark.** For the integral $I_9$, we can use the following recurrence formula

$$J_n := \int \cos^n x\, dx = \frac{\sin x \cos^{n-1} x}{n} + \frac{n - 1}{n}J_{n-2}; \quad n = 2, 3, \ldots.$$

Where $I_9 = J_6$.

$$\boxed{I_{10} = \int \frac{dx}{x^4 + 2x^2 + 1}}$$

**Solution 1.**

For the given integral we can write the following

$$I_{10} = \int \frac{dx}{(1 + x^2)^2} = \int \frac{1 + x^2 - x^2}{(1 + x^2)^2} dx = \int \frac{dx}{1 + x^2} - \underbrace{\int \frac{x^2}{(1 + x^2)^2} dx}_{:= J}.$$

For the integral $J$, we can write

$$J = \int \frac{x^2}{(1 + x^2)^2} dx = \int x \frac{x}{(1 + x^2)^2} dx.$$

By using the integration by parts, let us assume

$$u = x \implies du = dx,$$

$$dv = \frac{x}{(1+x^2)^2} dx \implies v = \int x(x^2+1)^{-2} dx = -\frac{1}{2(x^2+1)}.$$

Thus, we have

$$J = -\frac{x}{2(x^2+1)} + \frac{1}{2} \int \frac{dx}{1+x^2}.$$

Therefore, for the integral $I_{10}$, we have

$$I_{10} = \int \frac{dx}{1+x^2} + \frac{x}{2(x^2+1)} - \frac{1}{2} \int \frac{dx}{1+x^2} = \frac{1}{2} \int \frac{dx}{1+x^2} + \frac{x}{2(1+x^2)}$$

$$= \frac{1}{2} \arctan x + \frac{x}{2(1+x^2)} + C,$$

where $C \in \mathbb{R}$.

## Solution 2.

For the given integral we can write the following

$$I_{10} = \int \frac{dx}{(1+x^2)^2}.$$

Now, let us use the substitution $x = \tan t$, so that $dx = dt/\cos^2 t$. This gives the following

$$I_{10} = \int \frac{1}{(1+\tan^2 t)^2} \frac{dt}{\cos^2 t} = \int \cos^2 t \, dt = \frac{1}{2} \int (1+\cos(2t)) \, dt$$

$$= \frac{1}{2} \left( t + \frac{1}{2} \sin(2t) \right) + C = \frac{1}{2} \left( \arctan x + \frac{1}{2} \sin(2 \arctan x) \right) + C$$

$$= \frac{1}{2} \arctan x + \frac{1}{2} \sin(\arctan x) \cos(\arctan x) + C$$

$$= \frac{1}{2} \arctan x + \frac{1}{2} \frac{x}{\sqrt{1+x^2}} \frac{1}{\sqrt{1+x^2}} + C = \frac{1}{2} \arctan x + \frac{x}{2(1+x^2)} + C,$$

where $C \in \mathbb{R}$.

For the formulas of $\sin(\arctan x)$ and $\cos(\arctan x)$, see Subsec. 1.5.1.

$$\boxed{I_{11} = \int \cos(\ln x) dx}$$

## Solution 1.

By using the integration by parts, let us assume

$$u = \cos(\ln x) \implies du = -\frac{\sin(\ln x)}{x} dx, \quad dv = dx \implies v = x.$$

Thus, we have

$$I_{11} = x\cos(\ln x) + \underbrace{\int \sin(\ln x)dx}_{:=J}.$$

For the integral $J$, in a similar way, we assume

$$u = \sin(\ln x) \implies du = \frac{\cos(\ln x)}{x}dx, \quad dv = dx \implies v = x.$$

Thus we have

$$I_{11} = x\cos(\ln x) + x\sin(\ln x) - \underbrace{\int \cos(\ln x)dx}_{I_{11}}.$$

and

$$2I_{11} = x(\cos(\ln x) + \sin(\ln x)) + C_1,$$

where $C_1 \in \mathbb{R}$. Therefore,

$$I_{11} = \frac{x}{2}(\cos(\ln x) + \sin(\ln x)) + C,$$

where $C = C_1/2 \in \mathbb{R}$.

**Solution 2.**

Let us use the substitution $\ln x = t$, so that $x = e^t$ and $dx = e^t dt$. This gives the following

$$I_{11} = \int e^t \cos t\, dt = \frac{e^t}{2}(\cos t + \sin t) = \frac{x}{2}(\cos(\ln x) + \sin(\ln x)) + C,$$

where we used the formula (1.1), with $a = b = 1$.

$$\boxed{I_{12} = \int \frac{dx}{\cos x}}$$

**Solution.**

For the given integral we can write the following

$$I_{12} = \int \sec x\, dx = \int \sec x \frac{\sec x + \tan x}{\sec x + \tan x}dx = \int \frac{\sec^2 x + \sec x \tan x}{\sec x + \tan x}dx.$$

But

$$(\sec x + \tan x)' = \sec x \tan x + \sec^2 x.$$

Therefore,

$$I_{12} = \int \sec x\, dx = \ln|\sec x + \tan x| + C,$$

where $C \in \mathbb{R}$.

**Remark.** For the integral $I_{12}$, we can simply and directly write (see Subsec. 1.1.1)

$$\int \frac{dx}{\cos x} = \int \sec x dx = \ln |\sec x + \tan x| + C = \ln \left| \tan \left( \frac{\pi}{4} + \frac{x}{2} \right) \right| + C.$$

where $C \in \mathbb{R}$.

$$\boxed{I_{13} = \int \frac{dx}{9 \cos^2 x + 4 \sin^2 x}}$$

**Solution.**

For the given integral we can write the following

$$I_{13} = \int \frac{\frac{dx}{\cos^2 x}}{9 + 4 \frac{\sin^2 x}{\cos^2 x}} = \int \frac{1}{4 \left( \frac{9}{4} + \tan^2 x \right)} \frac{dx}{\cos^2 x}.$$

For the last integral, let us use the substitution $\tan x = t$, so that $dx / \cos^2 x = dt$. This gives the following

$$I_{13} = \frac{1}{4} \int \frac{dt}{(9/4) + t^2} = \frac{1}{4} \frac{2}{3} \arctan \left( \frac{2t}{3} \right) + C = \frac{1}{6} \arctan \left( \frac{2 \tan x}{3} \right) + C,$$

where $C \in \mathbb{R}$.

$$\boxed{I_{14} = \int \frac{dx}{x^2 (x^4 + 1)^{3/4}}}$$

**Solution.**

For the given integral we can write the following

$$I_{14} = \int \frac{dx}{x^2 \left( x^4 \right)^{3/4} \left( 1 + \frac{1}{x^4} \right)^{3/4}} = \int \frac{dx}{x^5 \left( 1 + \frac{1}{x^4} \right)^{3/4}}.$$

For the last integral, let us use the substitution $1 + \frac{1}{x^4} = t$, so that $\frac{dx}{x^5} = -\frac{dt}{4}$. This gives the following

$$I_{14} = -\frac{1}{4} \int t^{-3/4} dt = t^{1/4} + C = -\sqrt[4]{1 + \frac{1}{x^4}} + C = -\frac{\sqrt[4]{1 + x^4}}{x} + C,$$

where $C \in \mathbb{R}$.

In our solution, we assumed that $x > 0$, so $\sqrt[4]{x^4} = |x| = x$ (Be attentive when the integral is definite!).

$$I_{15} = \int_0^\pi \cos x \cos(3x) \cos(5x) dx$$

**Solution 1.**

By using the King property of integration (1.13), we find

$$I_{15} = \int_0^\pi \cos(\pi - x) \cos(3\pi - 3x) \cos(5\pi - 5x) dx$$

$$= \int_0^\pi (-\cos x)(-\cos(3x))(-\cos(5x)) dx$$

$$= -\underbrace{\int_0^\pi \cos x \cos(3x) \cos(5x) dx}_{I_{15}}.$$

Therefore,

$$2I_{15} = 0 \quad \Longrightarrow \quad I_{15} = 0.$$

**Solution 2.**

For the given integral we can write the following

$$I_{15} = \frac{1}{2} \int_0^\pi (\cos(4x) + \cos x) \cos(5x) dx$$

$$= \frac{1}{2} \int_0^\pi (\cos(4x) \cos(5x) + \cos x \cos(5x)) \, dx$$

$$= \frac{1}{2} \int_0^\pi \left( \frac{1}{2}(\cos(9x) + \cos x) + \frac{1}{2}(\cos(6x) + \cos(4x)) \right) dx$$

$$= \frac{1}{4} \int_0^\pi (\cos(9x) + \cos x + \cos(6x) + \cos(4x)) \, dx$$

$$= \frac{1}{4} \left[ \frac{1}{9} \sin(9x) + \sin x + \frac{1}{6} \sin(6x) + \frac{1}{4} \sin(4x) \right]_0^\pi = 0.$$

$$I_{16} = \int \left( \frac{1}{\ln x} + \ln(\ln x) \right) dx$$

**Solution 1.**

Note that

$$(x \ln(\ln x))' = \ln(\ln x) + \frac{1}{\ln x}.$$

Thus, we have

$$I_{16} = \int \left( \frac{1}{\ln x} + \ln(\ln x) \right) dx = \int (x \ln(\ln x))' \, dx = x \ln(\ln x) + C,$$

where $C \in \mathbb{R}$.

**Solution 2.**

We have

$$I_{16} = \int \left( \frac{1}{\ln x} + \ln(\ln x) \right) dx = \int \frac{dx}{\ln x} + \underbrace{\int \ln(\ln x) dx}_{:=J} \, .$$

For the integral $J$, by using the integration by parts, let us assume

$$u = \ln(\ln x) \implies du = \frac{dx}{x \ln x}, \quad dv = dx \implies v = x.$$

Thus, we have

$$J = x \ln(\ln x) - \int x \frac{dx}{x \ln x} = x \ln(\ln x) - \int \frac{dx}{\ln x}.$$

Therefore, for the given integral $I_{16}$, we find

$$I_{16} = \int \frac{dx}{\ln x} + x \ln(\ln x) - \int \frac{dx}{\ln x} = x \ln(\ln x) + C,$$

where $C \in \mathbb{R}$.

$$\boxed{I_{17} = \int \frac{dx}{2 + e^x}}$$

**Solution.**

For the given integral we can write the following

$$I_{17} = \int \frac{dx}{e^x \left( 1 + 2e^{-x} \right)} = \int \frac{e^{-x}}{1 + 2e^{-x}} dx.$$

For the last integral, let us use the substitution $e^{-x} = t$, so that $-e^{-x} dx = dt$. This gives the following

$$I_{17} = -\int \frac{dt}{1 + 2t} = -\frac{1}{2} \int \frac{2}{1 + 2t} dt = -\frac{1}{2} \ln |1 + 2t| + C$$

$$= -\frac{1}{2} \ln |1 + 2e^{-x}| + C = -\frac{1}{2} \left( \ln \left( 2 + e^x \right) - x \right) + C,$$

where $C \in \mathbb{R}$.

$$\boxed{I_{18} = \int \sqrt{\frac{x}{1 - x^3}} dx}$$

**Solution 1.**

For the given integral we can write the following

$$I_{18} = \int \frac{\sqrt{x}}{\sqrt{1-x^3}} dx = \int \frac{\sqrt{x}}{x^{3/2}\sqrt{x^{-3}-1}} dx.$$

Let us use the substitution $x^{3/2} = t$ (note that $t > 0$, because $x > 0$), so that $x = t^{2/3}$. Thus, $dx = \frac{2}{3} t^{-1/3} dt$. This gives the following

$$I_{18} = \frac{2}{3} \int \frac{t^{1/3} t^{-1/3}}{t\sqrt{t^{-2}-1}} dt = \frac{2}{3} \int \frac{dt}{t\sqrt{\frac{1-t^2}{t^2}}} = \frac{2}{3} \int \frac{dt}{\sqrt{1-t^2}} = \frac{2}{3} \arcsin t + C$$

$$= \frac{2}{3} \arcsin(x^{3/2}) + C,$$

where $C \in \mathbb{R}$.

**Solution 2.**

For the given integral we can write the following

$$I_{18} = \int x^{1/2} (1-x^3)^{-1/2} dx.$$

This is a binomial integral, from the form $\int x^m (ax^n + b)^p dx$, where $m = 1/2, n = 3, p = -1/2 \notin \mathbb{Z}$. But $\frac{m+1}{n} + p = 0 \in \mathbb{Z}$, therefore we can write

$$I_{18} = \int x^{1/2} (x^3)^{-1/2} \left(x^{-3} - 1\right)^{-1/2} dx = \int x^{-1} \left(x^{-3} - 1\right)^{-1/2} dx.$$

Now, let us use the substitution $x^{-3} - 1 = t^2$, so that $x = (1+t^2)^{-1/3}$. Thus $dx = -\frac{2}{3} t (1+t^2)^{-4/3} dt$. This gives the following

$$I_{18} = -\frac{2}{3} \int (1+t^2)^{1/3} (t^2)^{-1/2} (1+t^2)^{-4/3} dt$$

$$= -\frac{2}{3} \int (1+t^2)^{1/3} t^{-1} t (1+t^2)^{-4/3} dt$$

$$= -\frac{2}{3} \int \frac{dt}{1+t^2} = -\frac{2}{3} \arctan(t) + C$$

$$= -\frac{2}{3} \arctan\left(\sqrt{x^{-3}-1}\right) + C,$$

where $C \in \mathbb{R}$.

$$\boxed{I_{19} = \int \frac{4x}{1-x^4} dx}$$

**Solution.**

For the given integral we can write the following

$$I_{19} = 4 \int \frac{x}{1 - (x^2)^2} dx.$$

Now. let us use the substitution $x^2 = 1$, so that $2x dx = dt$. This gives the following

$$I_{19} = 2 \int \frac{dt}{1 - t^2} = \ln\left|\frac{1 + t}{1 - t}\right| + C = \ln\left|\frac{1 + x^2}{1 - x^2}\right| + C,$$

where $C \in \mathbb{R}$.

$$\boxed{I_{20} = \int x^x (1 + \ln x) dx}$$

**Solution.** (See (3.9), which is the integral $I_{13}$ in 2010 MIT Integration Bee).

For the given integral we can write the following

$$I_{20} = \int e^{x \ln x} (1 + \ln x) dx.$$

Now, let us use the substitution $x \ln x = t$, so that $(1 + \ln x) dx = dt$. This gives the following

$$I_{20} = \int e^t dt = e^t + C = e^{x \ln x} + C = x^x + C,$$

where $C \in \mathbb{R}$.

$$\boxed{I_{21} = \int_0^6 \sqrt{6x - x^2} dx}$$

**Solution 1.**

For the given integral we can write the following

$$I_{21} = \int_0^6 (6x)^{1/2} \sqrt{1 - \frac{x}{6}} dx = \sqrt{6} \int_0^6 x^{1/2} \left(1 - \frac{x}{6}\right)^{1/2} dx.$$

Now, let us use the substitution $x/6 = t$, so that $dx = 6dt$. If $x = 0$, then $t = 0$, and if $x = 6$, then $t = 1$. With these new values, we get the following

$$I_{21} = 6\sqrt{6} \int_0^6 (6t)^{1/2} (1 - t)^{1/2} dt = 36\sqrt{6} \int_0^6 t^{1/2} (1 - t)^{1/2} dt = 36\beta \left(3/2, 3/2\right).$$

Where $\beta(a,b) = \int_0^1 x^{a-1}(1-x)^{b-1}dx$ is the beta function (see 1.8.1 and 1.8.2). Remember that

$$\beta(a,b) = \int_0^1 x^{a-1}(1-x)^{b-1}dx = \frac{\Gamma(a)\Gamma(b)}{\Gamma(a+b)}, \quad \text{and} \quad \Gamma(1/2) = \sqrt{\pi}.$$

Therefore, we have

$$I_{21} = 36\frac{\Gamma(3/2)\Gamma(3/2)}{\Gamma(3)} = 36\frac{((1/2)\,\Gamma(1/2))^2}{2!} = \frac{18\pi}{4} = \frac{9\pi}{2}.$$

## Solution 2.

For the given integral we can write the following

$$I_{21} = \int_0^6 \sqrt{6x - x^2}dx = \int_0^6 \sqrt{-(x^2 - 6x + 9 - 9)}dx$$

$$= \int_0^6 \sqrt{9 - (x-3)^2}dx.$$

Let us use the substitution $t/3 = \sin y$, so that $dt = 3\cos y\,dy$. If $t = -3$, then $y = -\pi/2$, and if $t = 3$, then $y = \pi/2$. With these new values, we get the following

$$I_{21} = 9\int_{-\pi/2}^{\pi/2} \sqrt{1 - \sin^2 y}\,\cos y\,dy = 9\int_{-\pi/2}^{\pi/2} |\cos y|\cos y\,dy$$

$$= 18\int_0^{\pi/2} \cos^2 y\,dy = 9\int_0^{\pi/2} (1 + \cos(2y))dy$$

$$= 9\,[y - 2\sin(2y)]_0^{\pi/2} = \frac{9\pi}{2}.$$

$$\boxed{I_{22} = \int \sin(101x)\sin^{99} x\,dx}$$

## Solution.

For the given integral we can write the following

$$I_{22} = \int \sin(100x + x)\sin^{99} x\,dx$$

$$= \int (\sin(100x)\cos x + \cos(100x)\sin x)\sin^{99} x\,dx$$

$$= \int \sin(100x)\cos x\,\sin^{99} x\,dx + \underbrace{\int \cos(100x)\sin^{100} x\,dx}_{:=J}.$$

For the integral $J$, by using the integration by parts, let us assume

$$u = \sin^{100} x \implies du = 100 \sin^{99} x \cos x dx,$$

$$dv = \cos(100x)dx \implies v = \frac{1}{100} \sin(100x).$$

Thus, we have

$$J = \frac{1}{100} \sin(100x) \sin^{100} x - \int \sin(100x) \sin^{99} x \cos x dx.$$

Therefore, we have

$$I_{22} = \int \sin(100x) \sin^{99} x \cos x dx + \frac{1}{100} \sin^{100} x \sin(100x)$$
$$- \int \sin(100x) \sin^{99} x \cos x dx$$
$$= \frac{1}{100} \sin^{100} x \sin(100x) + C,$$

where $C \in \mathbb{R}$.

$$\boxed{I_{23} = \int x e^{e^{x^2} + x^2} dx}$$

**Solution.**

For the given integral we can write the following

$$I_{23} = \int x e^{e^{x^2}} e^{x^2} dx.$$

Let us use the substitution $x^2 = t$, so that $2x dx = dt$. This gives the following

$$I_{23} = \frac{1}{2} \int e^{e^t} e^t dt$$

Now, let us use the substitution $e^t = y$, so that $e^t dt = dy$. This gives the following

$$I_{23} = \frac{1}{2} \int e^y dy = \frac{1}{2} e^y + C = \frac{1}{2} e^{e^t} + C = \frac{1}{2} e^{e^{x^2}} + C,$$

where $C \in \mathbb{R}$.

$$\boxed{I_{24} = \int_0^1 \frac{x^3 - 3x^2 + 3x - 1}{x^4 + 4x^3 + 6x^2 + 4x + 1} dx}$$

**Solution.**

For the given integral we can write the following

$$I_{24} = \int_0^1 \underbrace{\frac{x^3 - 3x^2 + 3x - 1}{(x+1)^4}}_{:=f(x)}.$$

According to the method of partial fractions, we can write

$$f(x) = \frac{A}{x+1} + \frac{B}{(x+1)^2} + \frac{C}{(x+1)^3} + \frac{D}{(x+1)^4}.$$

Thus,

$$x^3 - 3x^2 + 3x - 1 = A(x+1)^3 + B(x+1)^2 + C(x+1) + D. \qquad (4.1)$$

By the corresponding between the sides of the equality (4.1), we find the following linear system of equations

$$\begin{cases} A = 1, \\ 3A + B = -3, \\ 3A + 2B + C = 3, \\ A + B + C + D = -1. \end{cases}$$

By solving this system, we find $A = 1$, $B = -6$, $C = 12$, $D = -8$. Therefore, for the given integral $I_{24}$, we get

$$I_{24} = \int_0^1 \left( \frac{1}{x+1} - \frac{6}{(x+1)^2} + \frac{12}{(x+1)^3} - \frac{8}{(x+1)^4} \right) dx$$

$$= \left[ \ln(x+1) + \frac{6}{x+1} - \frac{6}{(x+1)^2} + \frac{8}{3(x+1)^3} \right]_0^1$$

$$= \left( \ln 2 + 3 - \frac{3}{2} + \frac{1}{3} \right) - \frac{8}{3} = \ln 2 - \frac{5}{6}.$$

$$\boxed{I_{25} = \int \sqrt{\frac{1-x}{1+x}}\, dx}$$

**Solution.**

Let us use the substitution $\frac{1-x}{1+x} = t^2$, so that

$$1 - x = t^2 + xt^2 \implies x(1 + t^2) = 1 - t^2 \implies x = \frac{1 - t^2}{1 + t^2} \implies dx = \frac{-4t}{(1+t^2)^2}\, dt.$$

This gives the following

$$I_{25} = -4 \int t \frac{t}{(1+t^2)^2} \, dt.$$

By using the integration by parts, let us assume

$$u = t \implies du = dt, \quad dv = \frac{t}{(1+t^2)^2} dt \implies v = -\frac{1}{2(1+t^2)}.$$

Thus, we have

$$I_{25} = -4 \left( -\frac{t}{2(1+t^2)} + \frac{1}{2} \int \frac{dt}{1+t^2} \right) = \frac{2t}{1+t^2} - 2\arctan(t) + C$$

$$= \frac{2\sqrt{\frac{1-x}{1+x}}}{1 + \frac{1-x}{1+x}} - 2\arctan\left( \sqrt{\frac{1-x}{1+x}} \right) + C$$

$$= \frac{\sqrt{1-x}}{\sqrt{1+x}}(1+x) - \arctan\left( \sqrt{\frac{1-x}{1+x}} \right) + C$$

$$= \sqrt{1-x^2} - 2\arctan\left( \sqrt{\frac{1-x}{1+x}} \right) + C,$$

where $C \in \mathbb{R}$.

# Chapter 5

# The Solutions to the 2012 MIT Integration Bee, Qualifying Test

$$I_1 = \int \frac{dx}{\sqrt{x} - 1}$$

**Solution 1.**

Let us use the substitution $x = t^2$, so that $dx = 2tdt$. This gives the following

$$I_1 = 2 \int \frac{t}{t-1} dt = 2 \int \left( 1 + \frac{1}{t-1} \right) dt = 2t + 2 \ln|t - 1| + C$$
$$= 2\sqrt{x} + 2 \ln \left| \sqrt{x} - 1 \right| + C,$$

where $C \in \mathbb{R}$.

**Solution 2.**

Let us use the substitution $\sqrt{x} - 1 = t$, so that $x = (1 + t)^2$, and $dx = 2(1 + t)dt$. This gives the following

$$I_1 = 2 \int \frac{1+t}{t} dt = 2 \int \left( 1 + \frac{1}{t} \right) dt = 2t + 2 \ln|t| + C$$
$$= 2\sqrt{x} + 2 \ln \left| \sqrt{x} - 1 \right| + C,$$

where $C \in \mathbb{R}$.

$$I_2 = \int x^{1/4} \ln x dx$$

**Solution.**

By using the integration by parts, let us assume

$$u = \ln x \implies du = \frac{dx}{x}, \quad dv = x^{1/4}dx \implies v = \frac{4}{5}x^{5/4}.$$

Thus, we have

$$I_2 = \frac{4}{5}x^{5/4}\ln x - \frac{4}{5}\int x^{5/4}\frac{dx}{x} = \frac{4}{5}x^{5/4}\ln x - \frac{4}{5}\int x^{1/4}dx$$

$$= \frac{4}{5}x^{5/4}\ln x - \frac{16}{25}x^{5/4} + C = \frac{4}{5}x^{5/4}\left(\ln x - \frac{4}{5}\right) + C,$$

where $C \in \mathbb{R}$.

$$\boxed{I_3 = \int \frac{dx}{\left(1 + \sqrt{x}\right)\sqrt{x - x^2}}}$$

**Solution 1.**

For the given integral we can write the following

$$I_3 = \int \frac{dx}{\left(1 + \sqrt{x}\right)\sqrt{x}\sqrt{1 - x}}.$$

Let us use the substitution $1 + \sqrt{x} = t$ (note that $t > 0$, because $x > 0$), so that $x = (t - 1)^2$. Thus $\frac{dx}{2\sqrt{x}} = dt$. This gives the following

$$I_3 = 2\int \frac{dt}{t\sqrt{1 - (t - 1)^2}} = 2\int \frac{dt}{t\sqrt{2t - t^2}}.$$

Now, let us use the substitution $u = 1/t$ (note that $u > 0$, because $t > 0$), so that $dt = -du/u^2$. This gives the following

$$I_3 = -2\int \frac{du}{u^2\frac{1}{u}\sqrt{\frac{2}{u} - \frac{1}{u^2}}} = -2\int \frac{du}{\sqrt{2u - 1}} = -2\int (2u - 1)^{-1/2}du$$

$$= -2\sqrt{2u - 1} + C = -2\sqrt{\frac{2}{t} - 1} + C = -2\sqrt{\frac{2}{1 + \sqrt{x}} - 1} + C$$

$$= -2\sqrt{\frac{1 - \sqrt{x}}{1 + \sqrt{x}}} + C,$$

where $C \in \mathbb{R}$.

**Solution 2.**

Let us use the substitution $\sqrt{x} = t$ (note that $t > 0$, because $x > 0$), so that $x = t^2$. Thus $dx = 2t\,dt$. This gives the following

$$I_3 = 2\int \frac{t\,dt}{(1+t)\sqrt{t^2 - t^4}} = 2\int \frac{t\,dt}{(1+t)|t|\sqrt{1 - t^2}} = 2\int \frac{dt}{(1+t)\sqrt{1 - t^2}}.$$

Now, let us use the substitution $t = \sin u$, so that $dt = \cos u\,du$. This gives the following

$$I_3 = 2\int \frac{\cos u}{(1 + \sin u)\sqrt{1 - \sin^2 u}}\,du = 2\int \frac{du}{1 + \sin u} = 2\int \frac{1 - \sin u}{1 - \sin^2 u}\,du$$

$$= 2\int \frac{1 - \sin u}{\cos^2 u}\,du = 2\int \left(\frac{1}{\cos^2 u} - \frac{\sin u}{\cos^2 u}\right)\,du = 2\tan u - \frac{2}{\cos u} + C$$

$$= 2\tan(\arcsin t) - \frac{2}{\cos(\arcsin t)} + C = \frac{2t}{\sqrt{1 - t^2}} - \frac{2}{\sqrt{1 - t^2}} + C$$

$$= \frac{2(t - 1)}{\sqrt{1 - t^2}} + C = \frac{2(\sqrt{x} - 1)}{\sqrt{1 - x}} + C = -2\frac{1 - \sqrt{x}}{\sqrt{(1 - \sqrt{x})(1 + \sqrt{x})}} + C$$

$$= -2\sqrt{\frac{1 - \sqrt{x}}{1 + \sqrt{x}}} + C,$$

where $C \in \mathbb{R}$. In our solution, we assumed that $\cos u > 0$ (Be attentive when the integral is definite!).

Remember that if $\arcsin t = \theta$, then $t = \sin \theta$ and (see the inserted figure below)

$$\tan \theta = \tan(\arcsin t) = \frac{t}{\sqrt{1 - t^2}}, \qquad \cos \theta = \cos(\arcsin t) = \sqrt{1 - t^2}.$$

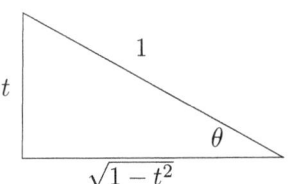

$$\boxed{I_4 = \int \frac{dx}{\sqrt{x}\left(\sqrt[4]{x} + 1\right)^{10}}}$$

**Solution.**

Let us use the substitution $x = t^4$, so that $dx = 4t^3\,dt$. This gives the following

$$I_4 = 4\int \frac{t^3}{t^2(1+t)^{-10}}dt = 4\int t(1+t)^{10}dt = 4\int (t+1-1)(1+t)^{-10}dt$$

$$= 4\int \left((1+t)^{-9} - (1+t)^{-10}\right)dt = 4\left(-\frac{1}{8(1+t)^8} + \frac{1}{9(1+t)^9}\right) + C$$

$$= -\frac{1}{2(1+t)^8} + \frac{4}{9(1+t)^9} + C = -\frac{1}{2\left(1+\sqrt[4]{x}\right)^8} + \frac{4}{9\left(1+\sqrt[4]{x}\right)^9} + C,$$

where $C \in \mathbb{R}$.

$$\boxed{I_5 = \int_0^1 \sin\left(\cos^{-1} x\right) dx}$$

**Solution.**

Let we set $y = \cos^{-1} x = \arccos x$, then $\cos y = x$ and $\sin y = \sin\left(\cos^{-1} x\right) = \sqrt{1-x^2}$. See the following figure

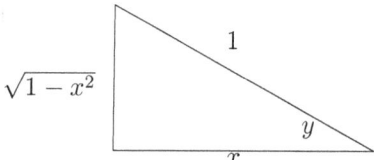

Therefore, for the given integral $I_5$, we have

$$I_5 = \int_0^1 \sin\left(\cos^{-1} x\right) dx = \int_0^1 \sqrt{1-x^2}dx.$$

Let us use the substitution $x = \sin t$, so that $dx = \cos t dt$. If $x = 0$, then $t = 0$, and if $x = 1$, then $t = \pi/2$. With these new values, we get the following

$$I_5 = \int_0^{\pi/2} \sqrt{1-\sin^2 t}\cos t\, dt = \int_0^{\pi/2} \cos^2 t\, dt = \frac{1}{2}\int_0^{\pi/2} (1+\cos(2t))\, dt$$

$$= \frac{1}{2}\left[t + \frac{1}{2}\sin(2t)\right]_0^{\pi/2} = \frac{1}{2}\frac{\pi}{2} = \frac{\pi}{4}.$$

$$\boxed{I_6 = \int \frac{dx}{\sqrt{1-4x-x^2}}}$$

**Solution.**

For the given integral we can write the following

$$I_6 = \int \frac{dx}{\sqrt{1-(x^2+4x)}} = \int \frac{dx}{\sqrt{5-(x+2)^2}} = \arcsin\left(\frac{x+2}{\sqrt{5}}\right) + C,$$

where $C \in \mathbb{R}$.

$$I_7 = \int_{1/4}^{1/2} \lfloor \ln \lfloor 1/x \rfloor \rfloor \, dx$$

**Solution.**

Let us use the substitution $1/x = t$, so that $dx = -dt/t^2$. If $x = 1/4$, then $t = 4$, and if $x = 1/2$, then $t = 2$. With these new values, we get the following

$$I_7 = \int_2^4 \frac{1}{t^2} \lfloor \ln \lfloor t \rfloor \rfloor \, dt = \int_2^3 \frac{1}{t^2} \lfloor \ln \lfloor t \rfloor \rfloor \, dt + \int_3^4 \frac{1}{t^2} \lfloor \ln \lfloor t \rfloor \rfloor \, dt.$$

But

$$\lfloor t \rfloor = 2, \ \forall t \in [2, 3[ \quad \Longrightarrow \quad \ln \lfloor t \rfloor = \ln 2, \ \forall t \in [2, 3[$$
$$\Longrightarrow \quad \lfloor \ln \lfloor t \rfloor \rfloor = 0, \ \forall t \in [2, 3[.$$

Similarly

$$\lfloor t \rfloor = 3, \ \forall t \in [3, 4[ \quad \Longrightarrow \quad \ln \lfloor t \rfloor = \ln 3, \ \forall t \in [3, 4[$$
$$\Longrightarrow \quad \lfloor \ln \lfloor t \rfloor \rfloor = 1, \ \forall t \in [3, 4[.$$

Therefore, we get the following

$$I_7 = \int_2^3 \frac{1}{t^2}(0)dt + \int_3^4 \frac{1}{t^2}(1)dt = \int_3^4 \frac{dt}{t^2} = \frac{1}{12}.$$

$$I_8 = \int_0^{\pi/2} \frac{dx}{1 + \sin x}$$

**Solution.**

Let us use the substitution $\tan\left(\frac{x}{2}\right) = t$, so that $dx = \frac{2}{1+t^2}dt$. If $x = 0$, then $t = 0$, and if $x = \frac{\pi}{2}$, then $t = 1$. With these new values, we get the following

$$I_8 = \int_0^1 \frac{\frac{2dt}{1+t^2}}{1 + \frac{2t}{1+t^2}} = 2 \int_0^1 \frac{dt}{t^2 + 2t + 1} = 2 \int_0^1 \frac{dt}{(t+1)^2}$$
$$= -2 \left[ \frac{1}{1+t} \right]_0^1 = -2 \left( \frac{1}{2} - 1 \right) = 1.$$

$$I_9 = \int_1^{2011} \frac{\sqrt{x}}{\sqrt{2012 - x} + \sqrt{x}} dx \qquad (5.1)$$

**Solution.**

By using the King property of integration (1.13), we find

$$
\begin{aligned}
I_9 &= \int_1^{2011} \frac{\sqrt{2012 - x}}{\sqrt{2012 - (2012 - x)} + \sqrt{2012 - x}}\,dx \\
&= \int_1^{2011} \frac{\sqrt{2012 - x}}{\sqrt{x} + \sqrt{2012 - x}}\,dx.
\end{aligned}
\tag{5.2}
$$

By adding (5.1) and (5.2), we find

$$
2I_9 = \int_1^{2011} \frac{\sqrt{x} + \sqrt{2012 - x}}{\sqrt{x} + \sqrt{2012 - x}}\,dx = \int_1^{2011} dx = 2010.
$$

Therefore,

$$
I_9 = 1005.
$$

$$
\boxed{I_{10} = \int \frac{x - 1}{(x + 1)\sqrt{x^3 + x^2 + x}}\,dx}
$$

**Solution.**

For the given integral we can write the following

$$
\begin{aligned}
I_{10} &= \int \frac{x^2 - 1}{(x + 1)^2\sqrt{x^3 + x^2 + x}}\,dx = \int \frac{x^2\left(1 - \frac{1}{x^2}\right)}{(x^2 + 2x + 1)\sqrt{x^2}\sqrt{x + 1 + \frac{1}{x}}}\,dx \\
&= \int \frac{x^2\left(1 - \frac{1}{x^2}\right)}{x^2\left(x + 2 + \frac{1}{x}\right)\sqrt{x + 1 + \frac{1}{x}}}\,dx = \int \frac{1 - \frac{1}{x^2}}{\left(x + \frac{1}{x} + 2\right)\sqrt{x + \frac{1}{x} + 1}}\,dx.
\end{aligned}
$$

Where we considered $x > 0$ (Be attentive when the integral is definite!). For the last integral, let us use the substitution $x + \frac{1}{x} + 1 = t^2$, so that $\left(1 - \frac{1}{x^2}\right)dx = 2t\,dt$. This gives the following

$$
I_{10} = 2 \int \frac{dt}{1 + t^2} = 2\arctan t + C = 2\arctan\left(\sqrt{x + \frac{1}{x} + 1}\right) + C,
$$

where $C \in \mathbb{R}$.

$$
\boxed{I_{11} = \int_{-1}^0 \frac{x^4 + 4x^3 + 6x^2 + 4x + 1}{x^3 - 3x^2 + 3x - 1}\,dx}
$$

**Solution 1.**

For the integrand in the given integral, according to the method of partial fractions, we can write

$$f(x) := \frac{x^4 + 4x^3 + 6x^2 + 4x + 1}{x^3 - 3x^2 + 3x - 1}$$

$$= \frac{x^4 + 4x^3 + 6x^2 + 4x + 1}{(x-1)^3}$$

$$= x + 7 + \frac{24x^2 - 16x + 8}{(x-1)^3}$$

$$= x + 7 + \frac{A}{x-1} + \frac{B}{(x-1)^2} + \frac{C}{(x-1)^3}.$$

Thus,

$$24x^2 - 16x + 8 = A(x-1)^2 + B(x-1) + C. \tag{5.3}$$

By the corresponding between the sides of the equality (5.3), we find the following linear system of equations

$$\begin{cases} A = 24, \\ 2A + B = -16, \\ A - B + C = 8. \end{cases}$$

By solving this system, we find $A = 24$, $B = 32$, $C = 16$. Therefore, for the given integral, we find

$$I_{11} = \int_{-1}^{0} \left( x + 7 + \frac{24}{x-1} + \frac{32}{(x-1)^2} + \frac{16}{(x-1)^3} \right) dx$$

$$= \left[ \frac{x^2}{2} + 7x + 24\ln|x-1| - \frac{32}{x-1} - \frac{8}{(x-1)^2} \right]_{-1}^{0}$$

$$= 24 - \left( 7 + \frac{1}{2} + 24\ln 2 \right) = \frac{33}{2} - 24\ln 2.$$

**Solution 2.**

The integrand in the given integral can be written as follows

$$f(x) := \frac{x^4 + 4x^3 + 6x^2 + 4x + 1}{x^3 - 3x^2 + 3x - 1} = \frac{(x+1)^4}{(x-1)^3}.$$

Let us use the substitution $x - 1 = t$, so that $dx = dt$. If $x = -1$, then $t = -2$, and if $x = 0$, then $t = -1$. With these new values, we get the following

$$I_{11} = \int_{-2}^{-1} \frac{(t+2)^4}{t^3} dt = \int_{-2}^{-1} \frac{t^4 + 8t^3 + 24t^2 + 32t + 16}{t^3}$$

$$= \int_{-2}^{-1} \left( t + 8 + \frac{24}{t} + 32t^{-2} + 16t^{-3} \right) dt$$

$$= \left[ \frac{t^2}{2} + 8t + 24 \ln |t| - \frac{32}{t} - \frac{8}{t^2} \right]_{-2}^{-1}$$

$$= \left( \frac{1}{2} + 16 \right) - (24 \ln 2) = \frac{33}{2} - 24 \ln 2.$$

$$I_{12} = \int \left( \cos x \ln x + \frac{\sin x}{x} \right) dx$$

**Solution 1.**

Note that

$$(\sin x \ln x)' = \cos x \ln x + \frac{\sin x}{x}.$$

Thus, we have

$$I_{12} = \int \left( \cos x \ln x + \frac{\sin x}{x} \right) dx = \int (\sin x \ln x)' \, dx = \sin x \ln x + C,$$

where $C \in \mathbb{R}$.

**Solution 2.**

For the given integral we can write the following

$$I_{12} = \int \cos x \ln x dx + \underbrace{\int \frac{\sin x}{x} dx}_{:=J} .$$

For the integral $J$, by using the integration by parts, let us assume

$$u = \sin x \Longrightarrow du = \cos x dx, \quad dv = \frac{dx}{x} \Longrightarrow v = \ln x.$$

Thus, we have

$$I_{12} = \int \cos x \ln x dx + \sin x \ln x - \int \cos x \ln x dx = \sin x \ln x + C,$$

where $C \in \mathbb{R}$.

$$I_{13} = \int \frac{dx}{x^3 - x}$$

**Solution.**

For the given integral we can write the following

$$I_{13} = \int \frac{\frac{dx}{x^3}}{1 - \frac{1}{x^2}}.$$

Let us use the substitution $\frac{1}{x^2} = t$, so that $\frac{dx}{x^3} = -\frac{dt}{2}$. This gives the following

$$I_{13} = \frac{1}{2}\int\frac{dt}{1-t} = \frac{1}{2}\int\frac{1}{t-1} = \frac{1}{2}\ln|t-1| + C = \frac{1}{2}\ln\left|1 - \frac{1}{x^2}\right| + C,$$

where $C \in \mathbb{R}$.

**Remark.** We can calculate the integral $I_{13}$ by the method of partial fractions, where we can write

$$\frac{1}{x^3 - x} = \frac{1}{x(x-1)(x+1)} = -\frac{1}{x} + \frac{1}{2(x-1)} + \frac{1}{2(x+1)}.$$

$$\boxed{I_{14} = \int_0^{1/2} \frac{x\sin^{-1}x}{\sqrt{1-x^2}}\,dx}$$

**Solution.**

Let us calculate the indefinite integral

$$J := \int\frac{x\sin^{-1}x}{\sqrt{1-x^2}}\,dx = \int\frac{x\arcsin x}{\sqrt{1-x^2}}\,dx.$$

By using the integration by parts, let us assume

$$u = \arcsin x \implies du = \frac{dx}{\sqrt{1-x^2}},$$

$$dv = \frac{x}{\sqrt{1-x^2}} \implies v = \int\frac{x}{\sqrt{1-x^2}}\,dx = -\sqrt{1-x^2}.$$

Thus, we have

$$J = -\sqrt{1-x^2}\arcsin x + \int\sqrt{1-x^2}\,\frac{dx}{\sqrt{1-x^2}}$$

$$= -\sqrt{1-x^2}\arcsin x + x + C,$$

where $C \in \mathbb{R}$. Therefore, for the given integral, we get

$$I_{14} = \left[-\sqrt{1-x^2}\arcsin x + x\right]_0^{1/2} = -\sqrt{\frac{3}{4}}\arcsin\left(\frac{1}{2}\right) + \frac{1}{2} = -\frac{\pi\sqrt{3}}{12} + \frac{1}{2}.$$

$$\boxed{I_{15} = \int_0^1 x(1-x)^{99}\,dx}$$

**Solution 1.**

For the given integral we can write the following

$$
I_{15} = -\int_0^1 (1 - x - 1)(1 - x)^{99} dx = -\int_0^1 \left((1 - x)^{100} - (1 - x)^{99}\right) dx
$$

$$
= \int_0^1 \left((1 - x)^{99} - (1 - x)^{100}\right) dx = \left[-\frac{1}{100}(1 - x)^{100} + \frac{1}{100}(1 - x)^{101}\right]_0^1
$$

$$
= -\left(-\frac{1}{100} + \frac{1}{101}\right) = \frac{1}{10100}.
$$

### Solution 2.

Let us use the substitution $1 - x = t$, so that $dx = -dt$. If $x = 0$, then $t = 1$, and if $x = 1$, then $t = 0$. With these new values, we get the following

$$
I_{15} = -\int_1^0 (1 + t)t^{99} dt = \int_0^1 \left(t^{99} + t^{100}\right) dt = \left[\frac{t^{100}}{100} + \frac{t^{101}}{101}\right]_0^1
$$

$$
= \frac{1}{100} - \frac{1}{101} = \frac{1}{10100}.
$$

### Solution 3.

By using the integration by parts, let us assume

$$
u = x \Longrightarrow du = dx, \quad dv = (1 - x)^{99} \Longrightarrow v = -\frac{1}{100}(1 - x)^{100}.
$$

Thus, we have

$$
I_{22} = \left[-\frac{x}{100}(1 - x)^{100}\right]_0^1 + \frac{1}{100}\int_0^1 (1 - x)^{100} dx
$$

$$
= \frac{1}{100}\left(-\frac{1}{101}\right)\left[(1 - x)^{101}\right]_0^1
$$

$$
= -\frac{1}{10100}(0 - 1) = \frac{1}{10100}.
$$

### Solution 4.

Let us calculate $I_{15}$ using the beta and gamma functions (see 1.8). We know that, for any $a > 0$ and $b > 0$, the beta function defined as follows

$$
\beta(a, b) = \int_0^1 x^{a-1}(1 - x)^{b-1} dx.
$$

Thus, for the integral $I_{15}$, we find $a - 1 = 1, b - 1 = 99$, and we can write

$$
I_{15} = \beta(2, 100) = \frac{\Gamma(2)\Gamma(100)}{\Gamma(102)} = \frac{(1!)(99!)}{101!} = \frac{99!}{(101)(100)(99!)} = \frac{1}{10100}.
$$

$$I_{16} = \int_0^{\pi/2} \frac{\sin(4x)}{\sin x} \, dx$$

## Solution.

For the given integral we can write the following

$$I_{16} = \int_0^{\pi/2} \frac{2\sin(2x)\cos(2x)}{\sin x} \, dx = \int_0^{\pi/2} \frac{4\sin x \cos x \cos(2x)}{\sin x} \, dx$$

$$= 4 \int_0^{\pi/2} \cos x \cos(2x) \, dx = 2 \int_0^{\pi/2} (\cos(3x) + \cos x) \, dx$$

$$= 2 \left[ \frac{1}{3}\sin(3x) + \sin x \right]_0^{\pi/2} = 2 \left( \frac{1}{3}\sin\left(\frac{3\pi}{2}\right) + \sin\left(\frac{\pi}{2}\right) \right) = \frac{4}{3}.$$

$$I_{17} = \int \frac{x^{-1/2}}{1 + x^{1/3}} \, dx$$

## Solution.

For the given integral we can write the following

$$I_{17} = \int \frac{dx}{x^{1/2}\left(1 + x^{1/3}\right)}.$$

Let us use the substitution $x = t^6$, so that $dx = 6t^5 dt$. This gives the following

$$I_{17} = 6 \int \frac{t^5}{t^3(1 + t^2)} \, dt = 6 \int \frac{t^2}{1 + t^2} \, dt = 6 \int \left(1 - \frac{1}{1 + t^2}\right) dt$$

$$= 6t - 6\arctan t + C = 6\sqrt[6]{x} - 6\arctan\left(\sqrt[6]{x}\right) + C,$$

where $C \in \mathbb{R}$.

$$I_{18} = \int \frac{dx}{\sqrt{2x^2 - 1}}$$

## Solution.

For the given integral we can write the following

$$I_{18} = \int \frac{dx}{\sqrt{(\sqrt{2}x)^2 - 1}}.$$

Let us use the substitution $\sqrt{2}x = t$, so that $\sqrt{2}dx = dt$. This gives the following

$$I_{18} = \frac{1}{\sqrt{2}} \int \frac{dt}{\sqrt{t^2 - 1}} = \frac{1}{\sqrt{2}} \ln \left| t + \sqrt{t^2 - 1} \right| + C$$
$$= \frac{1}{\sqrt{2}} \ln \left| \sqrt{2}x + \sqrt{2x^2 - 1} \right| + C,$$

where $C \in \mathbb{R}$.

$$I_{19} = \int \frac{dx}{\sqrt{e^x - 1}}$$

**Solution 1.**
Let us use the substitution $e^x = t$, so that $e^x dx = dt$, thus $dx = dt/t$. This gives the following

$$I_{19} = \int \frac{dt}{t\sqrt{t - 1}}.$$

Now, let us use the substitution $t - 1 = u^2$, so that $dt = 2u\,du$. This gives the following

$$I_{19} = 2 \int \frac{u}{(1 - u^2)u} du = 2 \int \frac{du}{1 + u^2} = 2 \arctan u + C = 2 \arctan \left( \sqrt{t - 1} \right)$$
$$= 2 \arctan \left( \sqrt{e^x - 1} \right) + C,$$

where $C \in \mathbb{R}$.
**Solution 2.**
Let us use the substitution $\sqrt{e^x - 1} = t$, so that $e^x = t^2 + 1 \implies x = \ln(1 + t^2)$, and $dx = \frac{2t}{1+t^2} dt$. This gives the following

$$I_{19} = \int \frac{1}{t} \frac{2t}{1 + t^2} dt = 2 \int \frac{dt}{1 + t^2} = 2 \arctan \left( \sqrt{e^x - 1} \right) + C,$$

where $C \in \mathbb{R}$.

$$I_{20} = \int \frac{x}{x^4 + 4} dx$$

**Solution.**
We have

$$I_{20} = \int \frac{x}{x^4 + 4} dx = \int \frac{x}{(x^2)^2 + 1} dx.$$

Let us use the substitution $x^2 = t$, so that $2x\,dx = dt$. This gives the following

$$I_{20} = \frac{1}{2} \int \frac{dt}{4 + t^2} = \frac{1}{2} \frac{1}{2} \arctan \left( \frac{t}{2} \right) + C = \frac{1}{4} \arctan \left( \frac{x^2}{2} \right) + C,$$

where $C \in \mathbb{R}$.

$$I_{21} = \int \frac{2dx}{(\cos x - \sin x)^2}$$

**Solution.**
We have

$$I_{21} = \int \frac{2dx}{(\cos x - \sin x)^2} = 2 \int \frac{dt}{\cos^2 x + \sin^2 x - 2 \sin x \cos x}$$

$$= 2 \int \frac{dx}{1 - \sin(2x)}.$$

For the last integral, let us use the substitution $\tan x = t$, so that $dx = \frac{dt}{1+t^2}$ and $\sin(2x) = \frac{2t}{1+t^2}$. This gives the following

$$I_{21} = 2 \int \frac{dt}{t^2 - 2t + 1} = 2 \int \frac{dt}{(t-1)^2} = -\frac{2}{t-1} + C = -\frac{2}{1 + \tan x} + C,$$

where $C \in \mathbb{R}$.

$$I_{22} = \int \frac{x \cosh x}{\sinh^2 x} dx$$

**Solution.**
By using the integration by parts, let us assume

$$u = x \Longrightarrow du = dx, \quad dv = \frac{\cosh x}{\sinh^2 x} \Longrightarrow v = \int \frac{\cosh x}{\sinh^2 x} dx = -\frac{1}{\sinh x}.$$

Thus, we have

$$I_{22} = -\frac{x}{\sinh x} + \int \frac{dx}{\sinh x} = -\frac{x}{\sinh x} + \int \frac{2}{e^x - e^{-x}} dx$$

$$= -\frac{x}{\sinh x} + 2 \int \frac{e^x}{(e^x)^2 - 1} dx.$$

For the last integral, let us use the substitution $e^x = t$, so that $e^x dx = dt$. This gives the following

$$I_{22} = -\frac{x}{\sinh x} + 2 \int \frac{dt}{1 - t^2} = -\frac{x}{\sinh x} + 2 - \ln \left| \frac{1+t}{1-t} \right| + C$$

$$= -\frac{x}{\sinh x} + 2 - \ln \left| \frac{1+e^x}{1-e^x} \right| + C,$$

where $C \in \mathbb{R}$.

$$I_{23} = \int_0^2 x^5 \sqrt{1 + x^3} dx$$

**Solution.**

For the given integral we can write the following

$$I_{23} = \int_0^2 x^5 \sqrt{1 + x^3} dx = \int_0^2 x^2 x^3 \sqrt{1 + x^3} dx.$$

Let us use the substitution $x^3 = t$, so that $3x^2 dx = dt$. If $x = 0$, then $t = 0$, and if $x = 2$, then $t = 8$. With these new values, we get the following

$$I_{23} = \frac{1}{3} \int_0^8 t(1 + t)^{1/2} dt = \frac{1}{3} \int_0^8 (t + 1 - 1)(1 + t)^{1/2} dt$$

$$= \frac{1}{3} \int_0^8 \left( (1 + t)^{3/2} - (1 + t)^{1/2} \right) dt = \frac{1}{3} \left[ \frac{2}{5}(1 + t)^{5/2} - \frac{2}{3}(1 + t)^{3/2} \right]_0^8$$

$$= \frac{1}{3} \left( \frac{2}{5}\sqrt{9^5} - \frac{2}{3}\sqrt{9^3} - \frac{2}{5} + \frac{2}{3} \right) = \frac{1192}{45}.$$

**Remark.** For the integral

$$\int t(1 + t)^{1/2} dt,$$

we can use the integration by parts. Let us assume

$$u = t \implies du = dt, \quad dv = (1 + t)^{1/2} dt \implies v = \frac{2}{3}(1 + t)^{3/2}.$$

Thus, we get

$$I_{23} = \frac{1}{3} \left( \left[ \frac{2}{3}t(1 + t)^{3/2} \right]_0^8 - \frac{2}{3} \int_0^8 (1 + t)^{3/2} dt \right)$$

$$= \frac{1}{3} \left( \left[ \frac{2}{3}t(1 + t)^{3/2} \right]_0^8 - \frac{2}{3} \int_0^8 (1 + t)^{3/2} dt \right)$$

$$= \frac{1}{3} \left( \frac{2}{3}8\sqrt{9^3} - \frac{2}{3}\frac{2}{5} \left[ (1 + t)^{5/2} \right]_0^8 \right)$$

$$= \frac{1}{3} \left( 144 - \frac{4}{15} \left( \sqrt{9^5} - 1 \right) \right) = \frac{1192}{45}.$$

$$I_{24} = \int_0^1 \frac{x^7 - 1}{\ln x} dx$$

**Solution.**

For the given integral we can write the following

$$I_{24} = \int_0^1 \frac{x^7 - 1}{\ln x} dx = \lim_{\varepsilon \to 0} \int_\varepsilon^{1-\varepsilon} \frac{x^7 - 1}{\ln x} dx.$$

Let us calculate the general form of the integral $I_{24}$, which is

$$I(a) = \int_0^1 \frac{x^a - 1}{\ln x} dx = \lim_{\varepsilon \to 0} \int_\varepsilon^{1-\varepsilon} \frac{x^a - 1}{\ln x} dx.$$

where $a \in \mathbb{R} \setminus \{-1\}$.

By Leibniz integral rule (1.15), we find

$$\frac{dI(a)}{da} = \lim_{\varepsilon \to 0} \int_\varepsilon^{1-\varepsilon} \frac{\partial}{\partial a} \left( \frac{x^a - 1}{\ln x} \right) dx = \lim_{\varepsilon \to 0} \int_\varepsilon^{1-\varepsilon} \left( \frac{1}{\ln x} x^a \ln x \right) dx$$

$$= \lim_{\varepsilon \to 0} \int_\varepsilon^{1-\varepsilon} x^a dx = \lim_{\varepsilon \to 0} \left( \frac{(1 - \varepsilon)^{a+1}}{a + 1} - \frac{\varepsilon^{a+1}}{a + 1} \right) = \frac{1}{a + 1}.$$

Thus

$$I(a) = \int \frac{da}{a + 1} = \ln |a + 1| + C,$$

where $C \in \mathbb{R}$. Now, let we set $a = 0$, then we find $I(0) = 0 = C$. Therefore, we get

$$I(a) = \int_0^1 \frac{x^a - 1}{\ln x} dx = \ln |a + 1|.$$

As a result, for the given integral $I_{24}$, we find

$$I_{24} = I(7) = \ln 8 = 3 \ln 2.$$

$$\boxed{I_{25} = \int \sqrt{\csc x - \sin x} dx}$$

**Solution.**

For the given integral we can write the following

$$I_{25} = \int \sqrt{\frac{1}{\sin x} - \sin x} dx = \int \sqrt{\frac{1 - \sin^2 x}{\sin x}} dx = \int \frac{|\cos x|}{\sqrt{\sin x}} dx$$

$$= \int \cos x (\sin x)^{-1/2} dx = 2\sqrt{\sin x} + C,$$

where $C \in \mathbb{R}$.

Note that, in our solution, we considered $\cos x > 0$ (Be attentive, when the integral is definite!).

# Chapter 6

# The Solutions to the 2013 MIT Integration Bee, Qualifying Test

$$I_1 = \int \left( \ln(x^2) - 2\ln(2x) \right) dx$$

**Solution.**

For the given integral we can write the following

$$I_1 = \int \left( 2\ln x - 2\ln(2x) \right) dx = 2 \int \left( \ln x - \ln(2x) \right) dx = 2 \int \ln \left( \frac{x}{2x} \right) dx$$

$$= 2\ln(1/2) \int dx = -\left( 2\ln 2 \right) x + C,$$

where $C \in \mathbb{R}$.

$$I_2 = \int_{-1}^{3} e^{|x|} dx$$

**Solution.**

For the given integral we can write the following

$$I_2 = \int_{-1}^{0} e^{-x} dx + \int_{0}^{3} e^{x} dx = \left[ -e^{-x} \right]_{-1}^{0} + \left[ e^{x} \right]_{0}^{3} = -\left( 1 - e \right) + \left( e^3 - 1 \right)$$

$$= e^3 + e - 2.$$

$$I_3 = \int \frac{\ln x \cos x - (\sin x / x)}{\ln^2 x} dx$$

**Solution 1.**

Note that

$$\left( \frac{\sin x}{\ln x} \right)' = \frac{\ln x \cos x - (\sin x / x)}{\ln^2 x}.$$

Thus, we have

$$I_3 = \int \frac{\ln x \cos x - (\sin x / x)}{\ln^2 x} dx = \int \left( \frac{\sin x}{\ln x} \right)' dx = \frac{\sin x}{\ln x} + C,$$

where $C \in \mathbb{R}$.

**Solution 2.**

For the given integral we can write the following

$$I_3 = \int \frac{\ln x \cos x}{\ln^2 x} dx - \int \frac{\sin x}{x \ln^2 x} dx = \underbrace{\int \frac{\cos x}{\ln x} dx}_{:=J} - \int \frac{\sin x}{x \ln^2 x} dx.$$

For the integral $J$, by using the integration by parts, let us assume

$$u = \frac{1}{\ln x} \implies du = -\frac{dx}{x \ln^2 x}, \quad dv = \cos x dx \implies v = \sin x.$$

Thus, we have

$$J = \frac{\sin x}{\ln x} + \int \frac{\sin x}{x \ln^2 x} dx.$$

Therefore, we get

$$I_3 = \frac{\sin x}{\ln x} + \int \frac{\sin x}{x \ln^2 x} dx - \int \frac{\sin x}{x \ln^2 x} dx = \frac{\sin x}{\ln x} + C,$$

where $C \in \mathbb{R}$.

$$I_4 = \int_1^{11} \left( x^3 - 3x^2 + 3x - 1 \right) dx$$

**Solution.**

For the given integral we can write the following

$$I_4 = \int_1^{11} (x - 1)^3 dx = \left[ \frac{(x - 1)^4}{4} \right]_1^{11} = \frac{1}{4} \left( 10^4 - 0 \right) = 2500.$$

$$I_5 = \int_0^2 \sqrt{12 - 3x^2} dx$$

**Solution.**

For the given integral we can write the following

$$I_5 = \int_0^2 \sqrt{12 - 3x^2} dx = \sqrt{12} \int_0^2 \sqrt{1 - (x/2)^2} dx.$$

Let us use the substitution $x/2 = \sin t$, so that $dx = 2 \cos t dt$. If $x = 0$, then $t = 0$, and if $x = 2$, then $t = \pi/2$. With these new values, we get the following

$$I_5 = 2\sqrt{3} \int_0^{\pi/2} \sqrt{1 - \sin^2 t}(2 \cos t) dt = 4\sqrt{3} \int_0^{\pi/2} \cos^2 t dt$$

$$= 4\sqrt{3} \int_0^{\pi/2} \frac{1 + \cos(2t)}{2} dt = 2\sqrt{3} \int_0^{\pi/2} (1 + \cos(2t)) dt$$

$$= 2\sqrt{3} \left[ t + \frac{1}{2} \sin(2t) \right]_0^{\pi/2} = 2\sqrt{3} \left( \frac{\pi}{2} \right) = \sqrt{3}\pi.$$

$$I_6 = \int_0^6 \left( x + (x - 3)^7 + \sin(x - 3) \right) dx$$

**Solution.**

For the given integral we can write the following

$$I_6 = \left[ \frac{x^2}{2} + \frac{(x - 3)^8}{8} - \cos(x - 3) \right]_0^6 = \left( 18 + \frac{3^8}{8} - \cos(3) \right) - \left( \frac{3^8}{8} - \cos(3) \right)$$

$$= 18.$$

$$I_7 = \int \sin x \sqrt{1 + \tan^2 x} dx$$

**Solution.**

For the given integral we can write the following

$$\int \sin x \sqrt{\frac{1}{\cos^2 x}} dx = \int \frac{\sin x}{\cos x} dx = -\ln|\cos x| + C,$$

where $C \in \mathbb{R}$.

Note that, in our solution, we considered $\cos x > 0$ (Be attentive, when the integral is definite!).

$$I_8 = \int \frac{x^5 - x^3 + x^2 - 1}{x^4 - x^3 + x - 1} dx$$

**Solution.**

For the given integral we can write the following

$$I_8 = \int \frac{x^3(x^2 - 1) + (x^2 - 1)}{x^3(x - 1) + (x - 1)} dx = \int \frac{(x^2 - 1)(x^3 + 1)}{(x - 1)(x^3 + 1)} dx$$
$$= \int \frac{(x - 1)(x + 1)}{x - 1} dx = \int (x + 1) dx = \frac{x^2}{2} + x + C,$$

where $C \in \mathbb{R}$.

$$I_9 = \int_0^1 \ln x \, dx$$

**Solution.**

For the given integral we can write the following

$$I_9 = \int_0^1 \ln x \, dx = \lim_{\varepsilon \to 0^+} \int_\varepsilon^{1-\varepsilon} \ln x \, dx.$$

By using the integration by parts, let us assume

$$u = \ln x \implies du = \frac{dx}{x}, \quad dv = dx \implies v = x.$$

Thus, we have

$$\int \ln x \, dx = x \ln x - \int dx = x \ln x - x + C,$$

where $C \in \mathbb{R}$. Thus for the given integral, we have

$$I_9 = \lim_{\varepsilon \to 0^+} [x \ln x - x]_\varepsilon^{1-\varepsilon}$$
$$= \lim_{\varepsilon \to 0^+} [((1 - \varepsilon) \ln(1 - \varepsilon) - (1 - \varepsilon)) - (\varepsilon \ln \varepsilon - \varepsilon)]$$
$$= -1.$$

$$I_{10} = \int \frac{dx}{1 - e^{-x}}$$

**Solution.**

For the given integral we can write the following

$$I_{10} = \int \frac{dx}{1 - e^{-x}} = \int \frac{e^x}{e^x - 1} dx = \ln|e^x - 1| + C,$$

where $C \in \mathbb{R}$.

$$I_{11} = \int_0^\pi \sin^2 x \cos^2 x\, dx$$

**Solution.**
For the given integral we can write the following

$$I_{11} = \int_0^\pi (\sin x \cos x)^2 dx = \int_0^\pi \left(\frac{1}{2}\sin(2x)\right)^2 dx = \frac{1}{4}\int_0^\pi \sin^2(2x)dx$$

$$= \frac{1}{4}\int_0^\pi \frac{1 - \cos(4x)}{2}dx = \frac{1}{8}\int_0^\pi (1 - \cos(4x))dx = \frac{1}{8}\left[x - \frac{1}{4}\sin(4x)\right]_0^\pi$$

$$= \frac{1}{8}(\pi - 0) = \frac{\pi}{8}.$$

$$I_{12} = \int_0^{441} \frac{\pi \sin\left(\pi\sqrt{x}\right)}{\sqrt{x}}dx$$

**Solution.**
Let us use the substitution $\pi\sqrt{x} = t$, so that $\frac{\pi}{\sqrt{x}}dx = 2dt$. If $x = 0$, then $t = 0$, and if $x = 441$, then $t = 21\pi$. With these new values, we get the following

$$I_{12} = 2\int_0^{21\pi} \sin t\, dt = -2\left[\cos t\right]_0^{21\pi} = -2(\cos(21\pi) - 1) = 4.$$

$$I_{13} = \int \tan^2 x\, dx$$

**Solution 1.**
For the given integral we can write the following

$$I_{13} = \int (1 + \tan^2 x - 1)\, dx = \int \left(\frac{1}{\cos^2 x} - 1\right) dx = \tan x - x + C,$$

where $C \in \mathbb{R}$.

**Solution 2.**

For the given integral we can write the following

$$I_{13} = \int \frac{\sin^2 x}{\cos^2 x} dx = \int \frac{1 - \cos^2 x}{\cos^2 x} dx = \int \left( \frac{1}{\cos^2 x} - 1 \right) dx = \tan x - x + C,$$

where $C \in \mathbb{R}$.

$$\boxed{I_{14} = \int_0^{256} (x - \lfloor x \rfloor)^2\, dx}$$

**Solution.**

Let $f(x) := (x - \lfloor x \rfloor)^2$. Then we find

$$\lfloor x \rfloor = 0, \quad \forall x \in [0, 1[ \implies f(x) = (x - 0)^2, \quad \forall x \in [0, 1[,$$
$$\lfloor x \rfloor = 1, \quad \forall x \in [1, 2[ \implies f(x) = (x - 1)^2, \quad \forall x \in [1, 2[,$$
$$\lfloor x \rfloor = 2, \quad \forall x \in [2, 3[ \implies f(x) = (x - 2)^2, \quad \forall x \in [2, 3[, \quad \ldots$$
$$\lfloor x \rfloor = 255, \quad \forall x \in [255, 256[ \implies f(x) = (x - 255)^2, \quad \forall x \in [255, 256[.$$

Thus we get:

$$I_{14} = \int_0^1 x^2 dx + \int_1^2 (x - 1)^2 dx + \int_2^3 (x - 2)^2 dx + \ldots + \int_{255}^{256} (x - 255)^2 dx$$

$$= \sum_{k=0}^{255} \left( \int_k^{k+1} (x - k)^2 dx \right) = \sum_{k=0}^{255} \left( \left[ \frac{(x - k)^3}{3} \right]_{x=k}^{x=k+1} \right)$$

$$= \sum_{k=0}^{255} \left( \frac{1}{3} - 0 \right) = \frac{256}{3}.$$

$$\boxed{I_{15} = \int e^{\sqrt[4]{x}}\, dx}$$

**Solution.**

Let us use the substitution $\sqrt[4]{x} = t$, so that $x = t^4$, and $dx = 4t^3 dt$. This gives the following

$$I_{15} = 4 \int t^3 e^t dt.$$

By using the integration by parts (tabular integration), we find

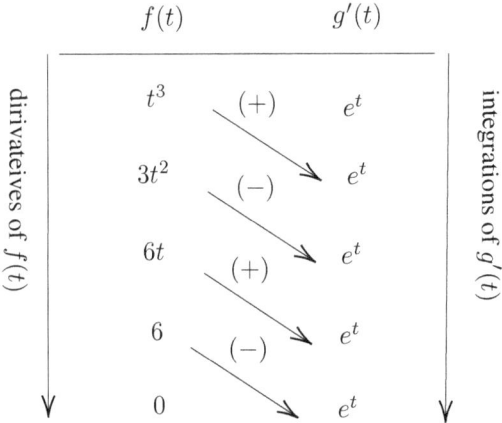

Thus,

$$I_{15} = 4\left(t^3 e^t - 3t^2 e^t + 6te^t - 6e^t\right) + C = 4e^t \left(t^3 - 3t^2 + 6t - 6\right) + C$$
$$= 4e^{\sqrt[4]{x}} \left(x^{3/4} - 3\sqrt{x} + 3\sqrt[4]{x} - 6\right) + C,$$

where $C \in \mathbb{R}$.

$$\boxed{I_{16} = \int \cos x \cot x dx}$$

**Solution.**

For the given integral we can write the following

$$I_{16} = \int \frac{\cos x}{\sin x} \cos x dx = \int \frac{\cos^2 x}{\sin x} dx = \int \frac{1 - \sin^2 x}{\sin x} dx$$
$$= \int \left(\frac{1}{\sin x} - \sin x\right) dx = \ln|\tan(x/2)| + \cos x + C,$$

where $C \in \mathbb{R}$.

$$\boxed{I_{17} = \int \left(2\ln x + \ln^2 x\right) dx}$$

**Solution 1.** (This integral was $I_2$ in the qualifying test of MIT Integration Bee 2011).
Note that

$$\left(x\ln^2 x\right)' = \ln^2 x + 2\ln x.$$

Thus, we have

$$I_2 = \int \left(2\ln x + \ln^2 x\right) dx = \int \left(x\ln^2 x\right)' dx = x\ln^2 x + C,$$

where $C \in \mathbb{R}$.

**Solution 2.**

For the given integral we can write the following

$$I_{17} = \int 2 \ln x dx + \underbrace{\int \ln^2 x dx}_{:=J}.$$

For the integral $J$, by using the integration by parts, let us assume

$$u = \ln^2 x \implies du = \frac{2 \ln x}{x} dx, \quad dv = dx \implies v = x.$$

Thus, we have

$$J = x \ln^2 x - 2 \int \ln x dx.$$

Therefore, we get

$$I_{17} = 2 \int \ln x dx + x \ln^2 x - 2 \int \ln x dx = x \ln^2 x + C,$$

where $C \in \mathbb{R}$.

$$\boxed{I_{18} = \int \frac{x^3}{1 + x^2} dx}$$

**Solution.**

For the given integral we can write the following

$$I_{18} = \int \left( x - \frac{x}{1 + x^2} \right) dx = \int \left( x - \frac{1}{2} \frac{2x}{1 + x^2} \right) dx$$
$$= \frac{x^2}{2} - \frac{1}{2} \ln(1 + x^2) + C,$$

where $C \in \mathbb{R}$.

$$\boxed{I_{19} = \int \frac{dx}{2 - 2x + x^2}}$$

**Solution.**

For the given integral we can write the following

$$I_{19} = \int \frac{dx}{x^2 - 2x + 1 + 1} = \int \frac{dx}{1 + (x - 1)^2} = \arctan(x - 1) + C,$$

where $C \in \mathbb{R}$.

$$I_{20} = \int \sin x \ln (\sin x)\, dx$$

**Solution.**

By using the integration by parts, let us assume

$$u = \ln (\sin x) \Longrightarrow du = \frac{\cos x}{\sin x}dx, \quad dv = \sin x dx \Longrightarrow v = -\cos x.$$

Thus, we have

$$I_{20} = -\cos x \ln (\sin x) + \int \frac{\cos^2 x}{\sin x}dx$$

$$= -\cos x \ln (\sin x) + \int \frac{1 - \sin^2 x}{\sin x}dx$$

$$= -\cos x \ln (\sin x) + \int \left(\frac{1}{\sin x} - \sin x\right) dx$$

$$= -\cos x \ln (\sin x) + \ln |\tan (x/2)| + \cos x + C,$$

where $C \in \mathbb{R}$.

$$I_{21} = \int \frac{x}{1 - x^4}dx$$

**Solution.**

Let us use the substitution $x^2 = t$, so that $2xdx = dt$. This gives the following

$$I_{21} = \frac{1}{2} \int \frac{dt}{1 - t^2} = \frac{1}{4} \ln \left|\frac{1+t}{1-t}\right| + C = \frac{1}{4} \ln \left|\frac{1+x^2}{1-x^2}\right| + C,$$

where $C \in \mathbb{R}$.

$$I_{22} = \int \sqrt{12 - 3x^2}dx$$

**Solution.**

For the given integral we can write the following

$$I_{22} = \int \sqrt{12 - 3x^2}dx = \sqrt{12} \int \sqrt{1 - (x/2)^2}dx.$$

Now, let us use the substitution $x/2 = \sin t$, so that $dx = 2\cos t\,dt$, and $t = \arcsin(x/2)$. This gives the following

$$I_{22} = 2\sqrt{3}\int \sqrt{1-\sin^2 t}(2\cos t)dt = 4\sqrt{3}\int \cos^2 t\,dt = 4\sqrt{3}\int \frac{1+\cos(2t)}{2}dt$$

$$= 2\sqrt{3}\int (1+\cos(2t))dt = 2\sqrt{3}\left(t + \frac{1}{2}\sin(2t)\right) + C$$

$$= 2\sqrt{3}\left(\arcsin\left(\frac{x}{2}\right) + \frac{1}{2}\sin\left(2\arcsin\left(\frac{x}{2}\right)\right)\right) + C$$

$$= 2\sqrt{3}\arcsin\left(\frac{x}{2}\right) + 2\sqrt{3}\sin\left(\arcsin\left(\frac{x}{2}\right)\right)\cos\left(\arcsin\left(\frac{x}{2}\right)\right) + C$$

$$= 2\sqrt{3}\arcsin\left(\frac{x}{2}\right) + \sqrt{3}x\sqrt{1-\left(\frac{x}{2}\right)^2} + C$$

$$= 2\sqrt{3}\arcsin\left(\frac{x}{2}\right) + \frac{\sqrt{3}}{2}x\sqrt{4-x^2} + C$$

$$= \frac{\sqrt{3}}{2}\left(x\sqrt{4-x^2} + 4\arcsin\left(\frac{x}{2}\right)\right) + C,$$

where $C \in \mathbb{R}$.

$$\boxed{I_{23} = \int \sec^5 x \tan^3 x\,dx}$$

**Solution.**

For the given integral we can write the following

$$I_{23} = \int \frac{1}{\cos^5 x}\frac{\sin^3 x}{\cos^3 x}dx = \int \cos^{-8}x\sin^2 x\sin x\,dx$$

$$= \int \cos^{-8}x(1-\cos^2 x)\sin x\,dx$$

$$= \int \left(\cos^{-8}x - \cos^{-6}x\right)\sin x\,dx.$$

For the last integral, let us use the substitution $\cos x = t$, so that $\sin x\,dx = -dt$. This gives the following

$$I_{23} = -\int \left(t^{-8} - t^{-6}\right)dt = -\left(-\frac{1}{7t^7} + \frac{1}{5t^5}\right) + C = \frac{1}{7t^7} - \frac{1}{5t^5} + C$$

$$= \frac{1}{7\cos^7 x} - \frac{1}{5\cos^5 x} + C = \frac{1}{35\cos^5 x}\left(\frac{5}{\cos^2 x} - 7\right) + C$$

$$= \frac{1}{35}\sec^5 x\left(5\sec^2 x - 7\right) + C,$$

where $C \in \mathbb{R}$.

$$I_{24} = \int_{-\pi/4}^{\pi/4} \frac{dx}{1 - \sin x}$$

## Solution 1.

For the given integral we can write the following

$$I_{24} = \int_{-\pi/4}^{\pi/4} \frac{1 + \sin x}{1 - \sin^2 x} dx = \int_{-\pi/4}^{\pi/4} \frac{1 + \sin x}{\cos^2 x} dx$$

$$= \int_{-\pi/4}^{\pi/4} \left( \frac{1}{\cos^2 x} + \sin x \cos^{-2} x \right) dx = \left[ \tan x + \frac{1}{\cos x} \right]_{-\pi/4}^{\pi/4}$$

$$= \left( \tan \left( \frac{\pi}{4} \right) + \frac{1}{\cos (\pi/4)} \right) - \left( \tan \left( -\frac{\pi}{4} \right) + \frac{1}{\cos (-\pi/4)} \right) = 2.$$

## Solution 2.

Let us use the substitution $\tan \left( \frac{x}{2} \right) = t$, so that $dx = \frac{2dt}{1+t^2}$, and $\sin x = \frac{2t}{1+t^2}$. If $x = -\frac{\pi}{4}$, then[1] $t = -\tan \left( \frac{\pi}{8} \right) = 1 - \sqrt{2}$, and if $x = \frac{\pi}{4}$, then $t = \tan \left( \frac{\pi}{8} \right) = \sqrt{2} - 1$. With these new values, we get the following

$$I_{24} = \int_{-\pi/8}^{\pi/8} \left( \frac{1}{1 - \frac{2t}{1+t^2}} \frac{2}{1 + t^2} \right) dt = 2 \int_{1-\sqrt{2}}^{\sqrt{2}-1} \frac{dt}{(t-1)^2} = -2 \left[ \frac{1}{t-1} \right]_{1-\sqrt{2}}^{\sqrt{2}-1}$$

$$= \frac{2}{1 - (\sqrt{2} - 1)} - \frac{2}{1 - (1 - \sqrt{2})} = \frac{2}{2 - \sqrt{2}} - \sqrt{2} = \frac{2(2 - \sqrt{2})}{2 - \sqrt{2}} = 2.$$

$$I_{25} = \int \frac{dx}{x\sqrt{x^2 - 2}}$$

## Solution 1.

Let us use the substitution $1/x = t$, so that $x = 1/t$, and $dx = -dt/t^2$. This gives the following

$$I_{25} = -\int \frac{dt}{t^2 \frac{1}{t} \sqrt{\frac{1}{t^2} - 2}} = -\int \frac{dt}{t \frac{\sqrt{1-2t^2}}{|t|}}.$$

If $x > 0$, then $t > 0$, and we get

$$I_{25} = -\int \frac{dt}{\sqrt{1 - 2t^2}} = -\int \frac{dt}{\sqrt{1 - (\sqrt{2}t)^2}}.$$

---

[1]To calculate $\tan \left( \frac{\pi}{8} \right)$. Remember that, $\tan(2x) = \frac{2 \tan x}{1 - \tan^2 x}$. So, by setting $x = \frac{\pi}{8}$, we find $\tan^2 \left( \frac{\pi}{8} \right) + 2 \tan \left( \frac{\pi}{8} \right) - 1 = 0$. By solving this quadratic equation we find the desired result, that is $\tan \left( \frac{\pi}{8} \right) = \sqrt{2} - 1$.

Now, for the last integral, let us use the substitution $\sqrt{2}\,t = u$, so that $dt = du/\sqrt{2}$. This gives the following

$$I_{25} = -\frac{1}{\sqrt{2}} \int \frac{du}{\sqrt{1-u^2}} = -\frac{1}{\sqrt{2}} \arcsin(u) + C = -\frac{1}{\sqrt{2}} \arcsin\left(\sqrt{2}\,t\right) + C$$

$$= -\frac{1}{\sqrt{2}} \arcsin\left(\frac{\sqrt{2}}{x}\right) + C,$$

where $C \in \mathbb{R}$, and $x > \sqrt{2}$.

In a similar way to the previous case, we find that if $x < 0$, then $t < 0$, and we have

$$I_{25} = \int \frac{dt}{\sqrt{1-2t^2}} = \frac{1}{\sqrt{2}} \arcsin\left(\sqrt{2}\,t\right) + C = \frac{1}{\sqrt{2}} = \frac{1}{\sqrt{2}} \arcsin\left(\frac{\sqrt{2}}{x}\right) + C,$$

where $C \in \mathbb{R}$, and $x < -\sqrt{2}$.

Therefore, for the given integral $I_{25}$, we get

$$I_{25} = \begin{cases} -\frac{1}{\sqrt{2}} \arcsin\left(\frac{\sqrt{2}}{x}\right) + C, & x > \sqrt{2}, \\ \frac{1}{\sqrt{2}} \arcsin\left(\frac{\sqrt{2}}{x}\right) + C, & x < -\sqrt{2}. \end{cases}$$

**Solution 2.**

For the given integral we can write the following

$$I_{25} = \int \frac{x}{x^2\sqrt{x^2-2}}\,dx.$$

Now, let us use the substitution $\sqrt{x^2-2} = t > 0$, so that $x^2 - 2 = t^2$, and $x\,dx = t\,dt$. This gives the following

$$I_{25} = \int \frac{t}{(t^2+2)|t|}\,dt = \int \frac{dt}{2+t^2} = \frac{1}{\sqrt{2}} \arctan\left(\frac{t}{\sqrt{2}}\right) + C$$

$$= \frac{1}{\sqrt{2}} \arctan\left(\frac{\sqrt{x^2-2}}{\sqrt{2}}\right) + C = -\frac{1}{\sqrt{2}} \arctan\left(\frac{\sqrt{2}}{\sqrt{x^2-2}}\right) + C',$$

where $C \in \mathbb{R}$ and $C' = C \pm (\pi/2) \in \mathbb{R}$.

Remember that

$$\arctan(1/x) + \arctan x = \pi/2, \quad \forall x \in ]0, \infty[,$$

$$\arctan(1/x) + \arctan x = -\pi/2, \quad \forall x \in ]-\infty, 0[.$$

# Chapter 7

# The Solutions to the 2014 MIT Integration Bee, Qualifying Test

$$I_1 = \int_1^e \ln(x^2)dx$$

**Solution.**

For the given integral we can write the following

$$I_1 = \int_1^e \ln(x^2)dx = 2\int_1^e \ln x\,dx.$$

By using the integration by parts, let us assume

$$u = \ln x \Longrightarrow du = \frac{dx}{x}, \quad dv = dx \Longrightarrow v = x.$$

Thus, we have

$$I_1 = 2\left([x\ln x]_1^e - \int_1^e dx\right) = 2(e - (e - 1)) = 2.$$

$$I_2 = \int_{-9}^9 \sin\left(\sqrt[3]{x}\right)dx$$

**Solution 1.**

Let $f(x) := \sin\left(\sqrt[3]{x}\right)$. Then, it holds that $f(-x) = -f(x)$, $\forall x \in [-9,9]$. Thus the given integral is an integral of an odd function. Therefore, $I_1 = 0$.

**Solution 2.**

Let us use the substitution $\sqrt[3]{x} = t$, so that $x = t^3$, and $dx = 3t^2 dt$. This gives the following

$$I := \int \sin\left(\sqrt[3]{x}\right) dx = 3 \int t^2 \sin t \, dt.$$

For the last integral, by using the integration by parts twice (at the first time we assume $u = t^2$ and $dv = \sin t \, dt$) we get the following

$$I = 3\left(-t^2 \cos t + 2t \sin t + 2 \cos t\right) + C$$
$$= -3\sqrt[3]{x^2} \cos\left(\sqrt[3]{x}\right) + 6\sqrt[3]{x} \sin\left(\sqrt[3]{x}\right) + 6 \cos\left(\sqrt[3]{x}\right) + C,$$

where $C \in \mathbb{R}$. Therefore for the given integral, we have

$$I_2 = \left[-3\sqrt[3]{x^2} \cos\left(\sqrt[3]{x}\right) + 6\sqrt[3]{x} \sin\left(\sqrt[3]{x}\right) + 6 \cos\left(\sqrt[3]{x}\right)\right]_{-9}^{9}$$
$$= \left(-3\sqrt[3]{81} \cos\left(\sqrt[3]{9}\right) + 6\sqrt[3]{9} \sin\left(\sqrt[3]{9}\right) + 6 \cos\left(\sqrt[3]{9}\right)\right)$$
$$- \left(-3\sqrt[3]{-81} \cos\left(\sqrt[3]{-9}\right) + 6\sqrt[3]{-9} \sin\left(\sqrt[3]{-9}\right) + 6 \cos\left(\sqrt[3]{-9}\right)\right)$$
$$= 0.$$

$$\boxed{I_3 = \int_0^\infty \left(\frac{d}{dx}\left(e^{1+x-x^2}\right)\right) dx}$$

**Solution.**

By applying the fundamental theorem of calculus (see Theorem 2), we find

$$I_3 = \lim_{b\to\infty} \int_0^b \left(\frac{d}{dx}\left(e^{1+x-x^2}\right)\right) dx = \lim_{b\to\infty}\left[e^{1+x-x^2}\right]_0^b$$
$$= \lim_{b\to\infty} e^{1+b-b^2} - e = 0 - e = -e.$$

$$\boxed{I_4 = \int_0^2 \sqrt{x + \sqrt{x + \sqrt{x + \sqrt{x + \ldots}}}} \, dx}$$

**Solution.**

Let

$$y := \sqrt{x + \sqrt{x + \sqrt{x + \ldots}}} \geqslant 0, \quad \forall x \in [0, 2].$$

Then

$$y^2 = x + \underbrace{\sqrt{x + \sqrt{x + \ldots}}}_{y} = x + y \quad \Longrightarrow \quad y^2 - y - x = 0.$$

By solving the quadratic equation $y^2 - y - x = 0$, for $y$, we find

$$y = \frac{1 - \sqrt{1 + 4x}}{2} < 0, \quad y = \frac{1 + \sqrt{1 + 4x}}{2} > 0, \quad \forall x \in [0, 2].$$

Therefore, for the given integral, we have

$$I_4 = \int_0^2 \left( \frac{1}{2} + \frac{1}{2}\sqrt{4x + 1} \right) dx = \left[ \frac{1}{2}x + \frac{1}{12}\sqrt{(4x + 1)^3} \right]_0^2$$

$$= 1 + \frac{\sqrt{9^3}}{12} - \frac{1}{12} = \frac{19}{6}.$$

$$\boxed{I_5 = \int \sqrt{x}\, e^{\sqrt{x}}\, dx}$$

**Solution.**
Let us use the substitution $\sqrt{x} = t$, so that $x = t^2$, and $dx = 2t\,dt$. This gives the following

$$I_5 = 2 \int t^2 e^t\, dt.$$

Now, by using the integration by parts twice (at the first time we assume $u = t^2$ and $dv = e^t\,dt$) we get the following

$$I_5 = 2\left(t^2 e^t - 2t e^t + 2e^t\right) + C = 2e^t\left(t^2 - 2t + 2\right) + C$$
$$= 2e^{\sqrt{x}}\left(x - 2\sqrt{x} + 2\right) + C,$$

where $C \in \mathbb{R}$.

$$\boxed{I_6 = \int \sin(2x)\cos(3x)\,dx}$$

**Solution.**
For the given integral we can write the following

$$I_6 = \frac{1}{2}\int \left(\sin(5x) + \sin(-x)\right) dx = \frac{1}{2}\int \left(\sin(5x) - \sin(x)\right) dx$$

$$= \frac{1}{2}\left(-\frac{1}{5}\cos(5x) + \cos x\right) + C = -\frac{1}{10}\cos(5x) + \frac{1}{2}\cos x + C,$$

where $C \in \mathbb{R}$.

$$\boxed{I_7 = \int_0^{2\pi} |1 + 2\sin x|\, dx}$$

**Solution.**

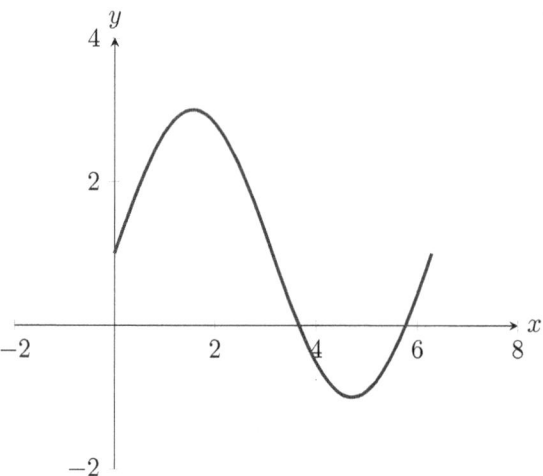

Figure 7.1:  The graph of $f(x) = 1 + 2\sin x$ in $[0, 2\pi]$.

At the first, let us find the solutions of the equation $1 + 2\sin x = 0$ in the interval $[0, 2\pi]$. For this, we have

$$1 + 2\sin x = 0 \implies \sin x = -\frac{1}{2} = \sin\left(-\frac{\pi}{6}\right).$$

Thus,

$$x = -\frac{\pi}{6} + 2\pi k, \quad \text{or} \quad x = \frac{7\pi}{6} + 2\pi k, \quad \text{where } k \in \mathbb{Z}.$$

From these sets of solutions, we find two solutions in the interval $[0, 2\pi]$, that are $\frac{7\pi}{6}, \frac{11\pi}{6}$ (see Fig 7.1). Therefore, for the given integral, we have

$$I_7 = \int_0^{7\pi/6} (1 + 2\sin x)dx - \int_{7\pi/6}^{11\pi/6} (1 + 2\sin x)dx + \int_{11\pi/6}^{2\pi} (1 + 2\sin x)dx$$

$$= [x - 2\cos x]_0^{7\pi/6} - [x - 2\cos x]_{7\pi/6}^{11\pi/6} + [x - 2\cos x]_{7\pi/6}^{2\pi}$$

$$= \left(\frac{7\pi}{6} - 2\cos\left(\frac{7\pi}{6}\right) + 2\right) - \left(\frac{11\pi}{6} - 2\cos\left(\frac{11\pi}{6}\right) - \frac{7\pi}{6} + 2\cos\left(\frac{7\pi}{6}\right)\right)$$

$$+ \left(2\pi - 2 - \frac{7\pi}{6} + 2\cos\left(\frac{7\pi}{6}\right)\right)$$

$$= 4\sqrt{3} + \frac{2\pi}{3}.$$

$$\boxed{I_8 = \int x(1 - x)^{2014} dx}$$

**Solution 1.**

For the given integral we can write the following

$$I_8 = -\int (1 - x - 1)(1 - x)^{2014} dx = -\int \left((1 - x)^{2015} - (1 - x)^{2014}\right) dx$$

$$= \frac{(1 - x)^{2016}}{2016} - \frac{(1 - x)^{2015}}{2015} + C,$$

where $C \in \mathbb{R}$.

**Solution 2.**

Let us use the substitution $1 - x = t$, so that $x = 1 - t$, and $dx = -dt$. This gives the following

$$I_8 = -\int (1 - t)t^{2014} dx = \int \left(t^{2015} - t^{2014}\right) dx = \frac{t^{2016}}{2016} - \frac{t^{2015}}{2015} + C$$

$$= \frac{(1 - x)^{2016}}{2016} - \frac{(1 - x)^{2015}}{2015} + C,$$

where $C \in \mathbb{R}$.

**Solution 3.**

By using the integration by parts, let us assume

$$u = x \Longrightarrow du = dx, \quad dv = (1 - x)^{2014} dx \Longrightarrow v = -\frac{1}{2015}(1 - x)^{2015}.$$

Thus, we have

$$I_9 = -\frac{x}{2015}(1 - x)^{2015} + \frac{1}{2015}\int (1 - x)^{2015} dx$$

$$= -\frac{x}{2015}(1 - x)^{2015} + \frac{1}{2015}\left(-\frac{1}{2016}\right)(1 - x)^{2016} + C$$

$$= -\frac{x}{2015}(1 - x)^{2015} - \frac{1}{4062240}(1 - x)^{2016} + C,$$

where $C \in \mathbb{R}$.

$$\boxed{I_9 = \int \operatorname{arsinh} x \, dx}$$

**Solution.**

By using the integration by parts, let us assume

$$u = \operatorname{arsinh} x \Longrightarrow du = \frac{dx}{\sqrt{1 + x^2}}, \quad dv = dx \Longrightarrow v = x.$$

Thus, we have

$$I_9 = x \operatorname{arsinh} x - \int \frac{x}{\sqrt{1+x^2}} dx = x \operatorname{arsinh} x - \frac{1}{2} \int \frac{2x}{\sqrt{1+x^2}} dx$$
$$= x \operatorname{arsinh} x - \sqrt{1+x^2} + C,$$

where $C \in \mathbb{R}$.

$$I_{10} = \int_{-1}^{0} \frac{x^2}{x-1} dx$$

**Solution.**

For the given integral we can write the following

$$I_{10} = \int_{-1}^{0} \left( x + 1 + \frac{1}{x-1} \right) dx = \left[ \frac{x^2}{2} + x + \ln|x-1| \right]_{-1}^{0}$$
$$= 0 - \left( \frac{1}{2} - 1 + \ln 2 \right) = \frac{1}{2} - \ln 2.$$

$$I_{11} = \int x \arctan x \, dx$$

**Solution.**

By using the integration by parts, let us assume

$$u = \arctan x \implies du = \frac{dx}{1+x^2}, \quad dv = x dx \implies v = \frac{x^2}{2}.$$

Thus, we have

$$I_{11} = \frac{x^2}{2} \arctan x - \frac{1}{2} \int \frac{x^2}{1+x^2} dx = \frac{x^2}{2} \arctan x - \frac{1}{2} \int \left( 1 - \frac{1}{1+x^2} \right) dx$$
$$= \frac{x^2}{2} \arctan x - \frac{x}{2} + \frac{1}{2} \arctan x + C,$$

where $C \in \mathbb{R}$.

$$I_{12} = \int \frac{dx}{x^2 - 15x - 2014}$$

**Solution.**

According to the method of partial fractions, we have

$$f(x) := \frac{1}{x^2 - 15x - 2014} = \frac{1}{(x-53)(x+38)} = \frac{A}{x-53} + \frac{B}{x+38},$$

where

$$A = \frac{1}{x+38}\Big|_{x=53} = \frac{1}{91}, \quad B = \frac{1}{x-53}\Big|_{x=-38} = -\frac{1}{91}.$$

Thus, we have

$$I_{12} = \int \left( \frac{1}{91(x-53)} - \frac{1}{91(x+38)} \right) dx$$

$$= \frac{1}{91}\ln|x-53| - \frac{1}{91}\ln|x+38| + C$$

$$= \frac{1}{91}\ln\left|\frac{x-53}{x+38}\right| + C,$$

where $C \in \mathbb{R}$.

$$I_{13} = \int e^x \left( \ln(1+x^2) - 2(1+x)\arctan x \right) dx$$

**Solution.**

By using the integration by parts, let us assume

$$u = \ln(1+x^2) - 2(1+x)\arctan x, \quad dv = e^x dx.$$

Then,

$$du = \left( \frac{2x}{1+x^2} - 2\left( \arctan x + \frac{1+x}{1+x^2} \right) \right) dx$$

$$= \left( -\frac{2}{1+x^2} - 2\arctan x \right) dx,$$

and $v = e^x$. Thus, we have

$$I_{13} = e^x \left( \ln(1+x^2) - 2(1+x)\arctan x \right) + 2\int e^x \left( \frac{1}{1+x^2} + \arctan x \right) dx$$

$$= e^x \left( \ln(1+x^2) - 2(1+x)\arctan x \right) + 2\int \left( \frac{e^x}{1+x^2} + e^x \arctan x \right) dx$$

$$= e^x \left( \ln(1+x^2) - 2(1+x)\arctan x \right) + 2\int (e^x \arctan x)' \, dx$$

$$= e^x \ln(1+x^2) - 2e^x \arctan x - 2xe^x \arctan x + 2e^x \arctan x + C$$

$$= e^x \left( \ln(1+x^2) - 2x\arctan x \right) + C,$$

where $C \in \mathbb{R}$.

**Remark.** For the integral

$$I = \int \left( \frac{e^x}{1+x^2} + e^x \arctan x \right) dx = \int \frac{e^x}{1+x^2}dx + \int e^x \arctan x\, dx,$$

by using the integration by parts, let us assume

$$u = \arctan x \implies du = \frac{dx}{1+x^2}, \quad dv = e^x dx \implies v = e^x.$$

Thus, we have

$$I = \int \frac{e^x}{1+x^2} dx + e^x \arctan x - \int \frac{e^x}{1+x^2} dx = e^x \arctan x + C.$$

$$\boxed{I_{14} = \int (\arcsin x)^2 \, dx}$$

**Solution.**

By using the integration by parts, let us assume

$$u = (\arcsin x)^2 \implies du = \frac{2\arcsin x}{\sqrt{1-x^2}} dx, \quad dv = dx \implies v = x.$$

Thus, we have

$$I_{14} = x(\arcsin x)^2 - 2 \int \frac{x}{\sqrt{1-x^2}} \arcsin x dx.$$

Now, for the last integral, by using the integration by parts, let us assume

$$u = \arcsin x \implies du = \frac{dx}{\sqrt{1-x^2}}, \quad dv = \frac{x}{\sqrt{1-x^2}} dx \implies v = -\sqrt{1-x^2}.$$

Thus, we have

$$I_{14} = x(\arcsin x)^2 - 2 \left( -\sqrt{1-x^2} \arcsin x + \int dx \right)$$
$$= x(\arcsin x)^2 + 2\sqrt{1-x^2} \arcsin x - 2x + C,$$

where $C \in \mathbb{R}$.

$$\boxed{I_{15} = \int \frac{\sqrt{x^2-1}}{x} dx}$$

**Solution 1.**

For the given integral, since $x \neq 0$, we can write the following

$$I_{15} = \int \frac{\sqrt{x^2-1}}{x} dx = \int \frac{\sqrt{x^2-1}}{x} \frac{x}{x} dx = \int \frac{\sqrt{x^2-1}}{x^2} x dx.$$

For the last integral, let us use the substitution $x^2 = t$, so that $2xdx = dt$. This gives the following

$$I_{15} = \frac{1}{2} \int \frac{\sqrt{t-1}}{t} dt.$$

Now, let us use the substitution $t - 1 = u^2$, so that $t = 1 + u^2$, and $dt = 2udu$. This gives the following

$$I_{15} = \frac{1}{2} \int \frac{u}{1+u^2} (2u) du = \int \frac{u^2}{1+u^2} du = \int \left(1 - \frac{1}{1+u^2}\right) du$$
$$= u - \arctan(u) + C = \sqrt{t-1} - \arctan\left(\sqrt{t-1}\right) + C$$
$$= \sqrt{x^2 - 1} - \arctan\left(\sqrt{x^2 - 1}\right) + C,$$

where $C \in \mathbb{R}$.

**Solution 2.**

Let us use the substitution $x = \frac{1}{\cos t}$, so that $dx = \frac{\sin t}{\cos^2 t} dt$. This gives the following

$$I_{15} = \int \frac{\sqrt{(1/\cos^2 t) - 1}}{1/\cos t} \frac{\sin t}{\cos^2 t} dt = \int \cos t \sqrt{\tan^2 t} \frac{\sin t}{\cos^2 t} dt$$
$$= \int \cos t \frac{\sin t}{\cos t} \frac{\sin t}{\cos^2 t} dt dt = \int \frac{\sin^2 t}{\cos^2 t} dt = \int \frac{1 - \cos^2 t}{\cos^2 t} dt$$
$$= \int \left(\frac{1}{\cos^2 t} - 1\right) dt = \tan t - t + C,$$

where $C \in \mathbb{R}$. But $x = 1/\cos t$, thus $\cos t = 1/x$ and $t = \arccos(1/x)$. Therefore, we get

$$I_{15} = \tan(\arccos(1/x)) - \arccos(1/x) + C = \sqrt{x^2 - 1} - \arccos(1/x) + C,$$

where $C \in \mathbb{R}$.

Note that, in our solution, we considered $\tan x > 0$ (Be attentive, when the integral is definite!).

**Remark.** By using the substitution $x = 1/\sin t$, we find

$$I_{15} = \sqrt{1 - x^2} + \arcsin(1/x) + C.$$

$$\boxed{I_{16} = \int x \sec^2(4x) dx}$$

**Solution.**

By using the integration by parts, let us assume

$$u = x \implies du = dx,$$

$$dv = \sec^2(4x)dx = \frac{dx}{\cos^2(4x)} \implies v = \int \frac{dx}{\cos^2(4x)} = \frac{1}{4}\tan(4x).$$

Thus, we have

$$I_{16} = \frac{1}{4}x\tan x - \frac{1}{4}\int \tan(4x)dx = \frac{1}{4}x\tan x - \frac{1}{4}\int \frac{\sin(4x)}{\cos(4x)}dx$$

$$= \frac{1}{4}x\tan x + \frac{1}{16}\int \frac{-4\sin(4x)}{\cos(4x)}dx = \frac{1}{4}x\tan x + \frac{1}{16}\ln|\cos(4x)| + C,$$

where $C \in \mathbb{R}$.

$$\boxed{I_{17} = \int \frac{2}{6 - 11x + 6x^2 - x^3}dx}$$

**Solution.**

For the given integral we can write the following

$$I_{17} = -2\int \frac{dx}{x^3 - 6x^2 + 11x - 6}.$$

According to the method of partial fractions, we have

$$f(x) := \frac{1}{x^3 - 6x^2 + 11x - 6} = \frac{1}{(x-1)(x-3)(x-2)}$$

$$= \frac{A}{x-1} + \frac{B}{x-3} + \frac{C}{x-2}.$$

where

$$A = \frac{1}{(x-3)(x-2)}\Big|_{x=1} = \frac{1}{2}, \quad B = \frac{1}{(x-1)(x-2)}\Big|_{x=3} = \frac{1}{2},$$

$$C = \frac{1}{(x-1)(x-3)}\Big|_{x=2} = -1.$$

Thus, we have

$$I_{17} = -2\int \left(\frac{1}{2(x-1)} + \frac{1}{2(x-3)} - \frac{1}{x-2}\right)dx$$

$$= -2\left(\frac{1}{2}\ln|x-1| + \frac{1}{2}\ln|x-3| - \ln|x-2|\right) + C'$$

$$= -\ln|x-1| - \ln|x-3| + 2\ln|x-2| + C'$$

$$= 2\ln|x-2| - \ln|(x-1)(x-3)| + C'$$

$$= \ln\left|\frac{(x-2)^2}{x^2 - 4x + 3}\right| + C' = \ln\left|\frac{x^2 - 4x + 4}{x^2 - 4x + 3}\right| + C',$$

where $C' \in \mathbb{R}$.

$$\boxed{I_{18} = \int_0^1 \frac{dx}{\lfloor 1 - \log_2(1 - x) \rfloor}}$$

**Solution.**

Let us find the values of $x$ in the interval $[0, 1[$, for which $1 - \log_2(1 - x) \in \mathbb{N} = \{1, 2, \ldots\}$. For this, we have the following equation by $x$

$$1 - \log_2(1 - x) = n; \quad n = 1, 2, \ldots$$

This means

$$\log_2(1 - x) = 1 - n \quad \Longrightarrow \quad x = 1 - 2^{1-n}; \quad n = 1, 2, \ldots$$

Therefore, when $x \in \{1 - 2^{1-n}; \ n = 1, 2, \ldots\}$, we have $1 - \log_2(1 - x) \in \mathbb{N}$. As a result, we find the following

$$\lfloor 1 - \log_2(1 - x) \rfloor = 1; \quad \forall x \in [0, 1/2[\,,$$
$$\lfloor 1 - \log_2(1 - x) \rfloor = 2; \quad \forall x \in [1/2, 3/4[\,,$$
$$\lfloor 1 - \log_2(1 - x) \rfloor = 3; \quad \forall x \in [3/4, 7/8[\,,$$
$$\lfloor 1 - \log_2(1 - x) \rfloor = 4; \quad \forall x \in [7/8, 15/16[\,, \quad \ldots$$
$$\lfloor 1 - \log_2(1 - x) \rfloor = n; \quad \forall x \in \left[1 - 2^{1-n}, 1 - 2^{-n}\right[\,.$$

Therefore, for the given integral, we get

$$I_{18} = \int_0^{1/2} dx + \int_{1/2}^{3/4} \frac{dx}{2} + \int_{3/4}^{7/8} \frac{dx}{3} + \int_{7/8}^{15/16} \frac{dx}{4} + \ldots + \int_{1-2^{1-n}}^{1-2^{-n}} \frac{dx}{n} + \ldots$$

$$= \sum_{n=1}^{\infty} \left( \int_{1-2^{1-n}}^{1-2^{-n}} \frac{dx}{n} \right)$$

$$= \sum_{n=1}^{\infty} \frac{1}{n} \left( 1 - 2^{-n} - 1 + 2^{1-n} \right)$$

$$= \sum_{n=1}^{\infty} \frac{1}{n} \left( -\frac{1}{2^n} + \frac{2}{2^n} \right)$$

$$= \sum_{n=1}^{\infty} \frac{1}{n 2^n} = \sum_{n=1}^{\infty} \frac{(1/2)^n}{n}.$$

Now, let us set

$$h(x) := \sum_{n=1}^{\infty} \frac{x^n}{n} = x + \frac{x^2}{2} + \frac{x^3}{3} + \ldots$$

This series is convergent for all $x$ in $[0, 1[$. Thus

$$h'(x) = \sum_{n=1}^{\infty} x^{n-1} = 1 + x + x^2 + \ldots = \frac{1}{1-x} \quad \forall x \in [0, 1[.$$

Therefore, $h(x) = -\ln|1 - x| + C$, where $C \in \mathbb{R}$ will be found. When $x = 0$, we find that $h(0) = 0 = C$. Thus $h(x) = -\ln|1 - x|$. As a result, for the given integral, we get

$$I_{18} = \int_0^1 \frac{dx}{[1 - \log_2(1 - x)]} = h(1/2) = \ln 2.$$

$$\boxed{I_{19} = \int_0^{1/\sqrt{3}} \sqrt{x + \sqrt{x^2 + 1}}\, dx}$$

**Solution 1.**

Let us use the substitution $x + \sqrt{x^2 + 1} = t$, then

$$\left(x + \sqrt{x^2 + 1}\right)^2 = t^2 \implies 2x^2 + 1 + 2x\sqrt{1 + x^2} = t^2$$

$$\implies 2x\underbrace{\left(x + \sqrt{1 + x^2}\right)}_{t} = t^2 - 1$$

$$\implies 2xt = t^2 - 1 \implies x = \frac{1}{2}\left(t - \frac{1}{t}\right).$$

If $x = 0$, then $t = 1$, and if $x = 1/\sqrt{3}$, then $t = \sqrt{3}$. With these new values, we get the following

$$I_{19} = \frac{1}{2}\int_1^{\sqrt{3}} \sqrt{t}\left(1 + \frac{1}{t^2}\right) dt = \frac{1}{2}\int_1^{\sqrt{3}} \left(t^{1/2} + t^{-3/2}\right) dt$$

$$= \frac{1}{2}\left[\frac{2}{3}t^{3/2} - 2t^{-1/2}\right]_1^{\sqrt{3}} = \frac{1}{2}\left[\frac{2}{3}\sqrt{t^3} - \frac{2}{\sqrt{t}}\right]_1^{\sqrt{3}}$$

$$= \frac{1}{2}\left(\frac{2}{3}\sqrt{3\sqrt{3}} - \frac{2}{\sqrt{\sqrt{3}}} - \left(\frac{2}{3} - 2\right)\right)$$

$$= \frac{1}{2}\left(\frac{2}{3}\sqrt{3}\sqrt{\sqrt{3}} - \frac{2}{\sqrt{\sqrt{3}}} + \frac{4}{3}\right) = \frac{1}{2}\left(\frac{2}{\sqrt{3}}\sqrt{\sqrt{3}} - \frac{2}{\sqrt{\sqrt{3}}} + \frac{4}{3}\right)$$

$$= \frac{1}{2}\left(\frac{2}{\sqrt{\sqrt{3}}} - \frac{2}{\sqrt{\sqrt{3}}} + \frac{4}{3}\right) = \frac{2}{3}.$$

**Solution 2.**

Let us use the substitution $x = \sinh t$, so that $dx = \cosh t dt$. If $x = 0$, then $t = 0$, and if $x = 1/\sqrt{3}$, then

$$\frac{e^t - e^{-t}}{2} = \frac{1}{\sqrt{3}} \quad \Longrightarrow \quad e^t - e^{-t} = \frac{2}{\sqrt{3}}, \Longrightarrow \quad \left(e^t\right)^2 - \frac{2}{\sqrt{3}} e^t - 1 = 0.$$

For solving this equation with respect to $e^t$, we have

$$\Delta = \left(\frac{2}{\sqrt{3}}\right)^2 - 4(1)(-1) = \frac{16}{3},$$

Therefore,

$$e^t = \frac{1}{2}\left(\frac{2}{\sqrt{3}} + \frac{4}{\sqrt{3}}\right) = \sqrt{3} \quad \Longrightarrow \quad t = \ln(\sqrt{3}).$$

With these new values, we get the following

$$I_{19} = \int_0^{\ln\sqrt{3}} \sqrt{\sinh t + \cosh t}\, \cosh t\, dt$$

$$= \int_0^{\ln\sqrt{3}} \sqrt{\frac{e^t - e^{-t}}{2} + \frac{e^t + e^{-t}}{2}}\, \frac{e^t + e^{-t}}{2}\, dt$$

$$= \frac{1}{2}\int_0^{\ln\sqrt{3}} e^{t/2}\left(e^t + e^{-t}\right) dt = \frac{1}{2}\int_0^{\ln\sqrt{3}} \left(e^{3t/2} + e^{-t/2}\right) dt$$

$$= \frac{1}{2}\left[\frac{2}{3} e^{3t/2} - 2e^{-t/2}\right]_0^{\ln\sqrt{3}} = \frac{1}{2}\left(\frac{2}{3} e^{(3\ln 3)/4} - 2e^{-(\ln 3)/4} - \left(\frac{2}{3} - 2\right)\right)$$

$$= \frac{1}{2}\left(\frac{2}{3} e^{\ln\left(3^{3/4}\right)} - 2e^{\ln\left(3^{-1/4}\right)} + \frac{4}{3}\right) = \frac{1}{2}\left(\frac{2}{3}\sqrt[4]{3^3} - \frac{2}{\sqrt[4]{3}} + \frac{4}{3}\right) = \frac{2}{3}.$$

$$\boxed{I_{20} = \int_0^{5\pi/2} \frac{dx}{2 + \cos x}}$$

**Solution.**

Let us use the substitution $\tan\left(\frac{x}{2}\right) = t$, so that $dx = \frac{2dt}{1+t^2}$, and $\cos x = \frac{1-t^2}{1+t^2}$. Note that the function $\varphi(x) = \tan(x/2)$ does not continuous at $\pi \in [0, 5\pi/2]$ (see Fig. 7.2). Therefore, we can write

$$I_{20} = \int_0^{\pi} \frac{dx}{2 + \cos x} + \int_{\pi}^{5\pi/2} \frac{dx}{2 + \cos x}$$

$$= \lim_{a\to\pi^-} \int_0^a \frac{dx}{2 + \cos x} + \lim_{b\to\pi^+} \int_b^{5\pi/2} \frac{dx}{2 + \cos x}.$$

Now, for the indefinite integral

$$I := \int \frac{dx}{2 + \cos x},$$

we have

$$I = \int \frac{\frac{2}{1+t^2}}{2 + \frac{1-t^2}{1+t^2}} \, dt = 2 \int \frac{dt}{3 + t^2} = \frac{2}{\sqrt{3}} \arctan \left( \frac{t}{\sqrt{3}} \right) + C$$

$$= \frac{2}{\sqrt{3}} \arctan \left( \frac{\tan(x/2)}{\sqrt{3}} \right) + C,$$

where $C \in \mathbb{R}$. Thus, for the given integral, we get

$$I_{20} = \frac{2}{\sqrt{3}} \lim_{a \to \pi^-} \left[ \arctan \left( \frac{\tan(x/2)}{\sqrt{3}} \right) \right]_0^a$$

$$+ \frac{2}{\sqrt{3}} \lim_{b \to \pi^+} \left[ \arctan \left( \frac{\tan(x/2)}{\sqrt{3}} \right) \right]_b^{5\pi/2}$$

$$= \frac{2}{\sqrt{3}} \left( \frac{\pi}{2} - 0 \right) + \frac{2}{\sqrt{3}} \left( \frac{\pi}{6} - \left( -\frac{\pi}{2} \right) \right) = \frac{7\pi}{3\sqrt{3}}.$$

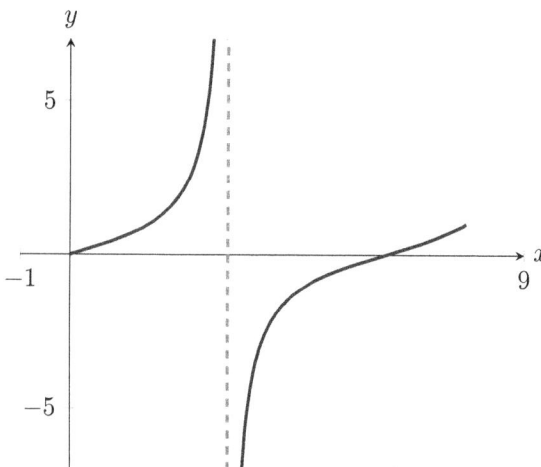

Figure 7.2: The graph of the function $\varphi(x) = \tan(x/2)$ in $[0, 5\pi/2]$. The dashed line is $x = \pi$.

**Remark.** [Be attentive!] Many readers may be thought to calculate the integral $I_{20}$, as follows.

By using the substitution $\tan\left(\frac{x}{2}\right) = t$, we find $dx = \frac{2dt}{1+t^2}$, and $\cos x = \frac{1-t^2}{1+t^2}$. If $x = 0$, then $t = 0$, and if $x = 5\pi/2$, then $t = 1$. With these new values, we get the following

$$I_{20} = \int_0^1 \frac{\frac{2}{1+t^2}}{2 + \frac{1-t^2}{1+t^2}} dt = 2 \int_0^1 \frac{dt}{3+t^2} = \left[\frac{2}{\sqrt{3}} \arctan\left(\frac{t}{\sqrt{3}}\right)\right]_0^1$$

$$= \frac{2}{\sqrt{3}}\left(\arctan\left(\frac{1}{\sqrt{3}}\right) - 0\right) = \frac{\pi}{3\sqrt{3}}.$$

This solution is not true, because the function $\varphi(x) = \tan(x/2)$ does not continuous at $\pi \in [0, 5\pi/2]$ (see Fig. 7.2).

# Chapter 8

# The Solutions to the 2015 MIT Integration Bee, Qualifying Test

$$I_1 = \int \left( \cos^4 x - \sin^4 x \right) dx$$

**Solution.**

For the given integral we can write the following

$$I_1 = \int \left( \cos^2 x - \sin^2 x \right) \left( \cos^2 x + \sin^2 x \right) dx = \int \left( \cos^2 x - \sin^2 x \right) dx$$

$$= \int \cos(2x) dx = \frac{1}{2} \sin(2x) + C,$$

where $C \in \mathbb{R}$.

$$I_2 = \int \frac{x}{\sqrt{2 + 4x}} dx$$

**Solution 1.**

For the given integral we can write the following

$$I_2 = \int x(2+4x)^{-1/2}dx = \frac{1}{4}\int (2+4x-2)(2+4x)^{-1/2}dx$$

$$= \frac{1}{4}\int \left((2+4x)^{1/2} - 2(2+4x)^{-1/2}\right)dx$$

$$= \frac{1}{4}\left(\frac{1}{6}(2+4x)^{3/2} - (2+4x)^{1/2}\right) + C$$

$$= \frac{1}{4}\left(\frac{1}{6}(2+4x)\sqrt{2+4x} - \sqrt{2+4x}\right) + C$$

$$= \frac{1}{4}\sqrt{2+4x}\left(-\frac{2}{3}+\frac{2}{3}x\right) + C = \frac{1}{3}(x-1)\sqrt{x+\frac{1}{2}} + C,$$

where $C \in \mathbb{R}$.

**Solution 2.**

Let us use the substitution $2+4x = t$, so that $x = (t-2)/4$, and $dx = dt/4$. This gives the following

$$I_2 = \frac{1}{4}\int \frac{t-2}{t^{1/2}}\frac{dt}{4} = \frac{1}{16}\int \left((t-2)t^{-1/2}\right)dt = \frac{1}{16}\int \left(t^{1/2} - 2t^{-1/2}\right)dt$$

$$= \frac{1}{16}\left(\frac{2}{3}t^{3/2} - 4t^{1/2}\right) + C = \frac{1}{24}\sqrt{(2+4x)^3} - \frac{1}{4}\sqrt{2+4x} + C$$

$$= \frac{1}{6}(x-1)\sqrt{2+4x} + C,$$

where $C \in \mathbb{R}$.

**Solution 3.**

By using the integration by parts, let us assume

$$u = x \implies du = dx,$$

$$dv = \frac{dx}{\sqrt{2+4x}} \implies v = \int (2+4x)^{-1/2}dx = \frac{1}{2}(2+4x)^{1/2}.$$

Thus, we have

$$I_2 = \frac{x}{2}(2+4x)^{\frac{1}{2}} - \frac{1}{2}\int (2+4x)^{\frac{1}{2}}dx = \frac{x}{2}(2+4x)^{\frac{1}{2}} - \frac{1}{2}\frac{1}{4}\frac{2}{3}(2+4x)^{\frac{3}{2}} + C$$

$$= \frac{x}{2}\sqrt{2+4x} - \frac{1}{12}(2+4x)^{3/2} + C = \frac{1}{2}\sqrt{2+4x}\left(x - \frac{1}{3}(1+2x)\right) + C$$

$$= \frac{1}{6}(x-1)\sqrt{2+4x} + C,$$

where $C \in \mathbb{R}$.

$$I_3 = \int_0^8 \frac{\cos\left(\sqrt{x}\right)}{\sqrt{x}}\,dx$$

**Solution.**

Let us use the substitution $\sqrt{x} = t$, so that $\frac{dx}{\sqrt{x}} = 2dt$. If $x = 0$, then $t = 0$, and if $x = 8$, then $t = 2\sqrt{2}$. With these new values, we get the following

$$I_3 = 2\int_0^{2\sqrt{2}} \cos t\,dt = 2\left[\sin t\right]_0^{2\sqrt{2}} = 2\sin(2\sqrt{2}).$$

$$I_4 = \int \sec x\,dx$$

**Solution 1.**

For the given integral we can write the following

$$I_4 = \int \frac{\sec x + \tan x}{\sec x + \tan x}\sec x\,dx = \int \frac{\sec^2 x + \sec x \tan x}{\sec x + \tan x}\,dx.$$

But

$$(\sec x + \tan x)' = \sec x \tan x + \sec^2 x.$$

Thus, we have

$$I_4 = \int \frac{(\sec x + \tan x)'}{\sec x + \tan x}\,dx = \ln|\sec x + \tan x| + C,$$

where $C \in \mathbb{R}$.

**Solution 2.**

Let us use the substitution $\tan\left(\frac{x}{2}\right) = t$, so that $dx = \frac{2dt}{1+t^2}$, and $\cos x = \frac{1-t^2}{1+t^2}$. This gives the following

$$I_4 = \int \frac{1+t^2}{1-t^2}\frac{2}{1+t^2}\,dt = 2\int \frac{dt}{1-t^2} = 2\frac{1}{2}\ln\left|\frac{1+t}{1-t}\right| + C$$

$$= \ln\left|\frac{1+\tan\left(x/2\right)}{1-\tan\left(x/2\right)}\right| + C = \ln\left|\frac{1+\sin\left(x/2\right)}{1-\cos\left(x/2\right)}\right| + C,$$

where $C \in \mathbb{R}$.

$$I_5 = \int_0^{\pi/2} \frac{e^{\sin x}}{\tan x \csc x}\,dx$$

**Solution.**

For the given integral we can write the following

$$I_5 = \int_0^{\pi/2} \frac{e^{\sin x}}{\tan x \csc x} dx = \int_0^{\pi/2} \frac{e^{\sin x}}{\frac{\sin x}{\cos x} \frac{1}{\sin x}} dx = \int_0^{\pi/2} \cos x e^{\sin x} dx$$

$$= \left[ e^{\sin x} \right]_0^{\pi/2} = e - 1.$$

$$I_6 = \int_1^e x \ln^2 x dx$$

**Solution.**

By using the integration by parts, let us assume

$$u = \ln^2 x \Longrightarrow du = \frac{2 \ln x}{x} dx, \quad dv = x dx \Longrightarrow v = \frac{x^2}{2}.$$

Thus, we have

$$I_6 = \left[ \frac{x^2}{2} \ln^2 x \right]_1^e - \int_1^e \frac{x^2}{2} \frac{2 \ln x}{x} dx = \left( \frac{e^2}{2} - 0 \right) - \int_1^e x \ln x dx$$

$$= \frac{e^2}{2} - \int_1^e x \ln x dx.$$

Now, for the last integral, by using the integration by parts, let us assume

$$u = \ln x \Longrightarrow du = \frac{dx}{x}, \quad dv = x dx \Longrightarrow v = \frac{x^2}{2}.$$

Thus, we have

$$I_6 = \frac{e^2}{2} - \left( \left[ \frac{x^2}{2} \ln x \right]_1^e - \frac{1}{2} \int_1^e x dx \right) = \frac{e^2}{2} - \left( \frac{e^2}{2} - 0 - \frac{1}{2} \left[ \frac{x^2}{2} \right]_1^e \right)$$

$$= \frac{1}{2} \left( \frac{e^2}{2} - \frac{1}{2} \right) = \frac{1}{4} \left( e^2 - 1 \right).$$

$$I_7 = \int \frac{dx}{5 + 4\sqrt{x} + x}$$

**Solution.**

Let us use the substitution $\sqrt{x} = t$, so that $x = t^2$, and $dx = 2t dt$. This gives the following

$$I_7 = \int \frac{2t}{t^2 + 4t + 5} dt = \int \frac{2t - 4 + 4}{t^2 + 4t + 5} dt$$

$$= \int \frac{2t + 4}{t^2 + 4t + 5} dt - 4 \int \frac{dt}{t^2 + 4t + 5}$$

$$= \int \frac{2t+4}{t^2+4t+5}dt - 4\int \frac{dt}{(t+2)^2+1}$$
$$= \ln \left|t^2+4t+5\right| - 4\arctan(t+2) + C$$
$$= \ln \left|x+4\sqrt{x}+5\right| - 4\arctan\left(\sqrt{x}+2\right) + C,$$

where $C \in \mathbb{R}$.

$$\boxed{I_8 = \int (2015)^x dx}$$

**Solution.**

By the fundamental rules of integration, We can directly write

$$I_8 = \int (2015)^x dx = \frac{(2015)^x}{\ln(2015)} + C,$$

where $C \in \mathbb{R}$.

$$\boxed{I_9 = \int_0^2 \frac{x}{(x-3)(x+5)^2}dx}$$

**Solution.**

According to the method of partial fractions, we have

$$f(x) := \frac{x}{(x+5)^2(x-3)} = \frac{A}{x+5} + \frac{B}{(x+5)^2} + \frac{C}{x-3}.$$

Thus,

$$x = A(x+5)(x-3) + B(x-3) + C(x+5)^2. \qquad (8.1)$$

By the corresponding between the sides of the equality (8.1), we find the following linear system of equations

$$\begin{cases} A + C = 0, \\ 2A + B + 10C = 1, \\ -15A - 3B + 25C = 0. \end{cases}$$

By solving this system, we find $A = -\frac{3}{64}$, $B = \frac{5}{8}$, $C = \frac{3}{64}$. Therefore, for the given integral, we find

$$I_9 = \int_0^2 \left(-\frac{3}{64(x+5)} + \frac{5}{8(x+5)^2} + \frac{3}{64(x-3)}\right) dx$$
$$= \left[-\frac{3}{64}\ln|x+5| - \frac{5}{8(x+5)} + \frac{3}{64}\ln|x-3|\right]_0^2$$
$$= \left[-\frac{3}{64}\ln\left|\frac{x+5}{x-3}\right| - \frac{5}{8(x+5)}\right]_0^2$$

$$= \left(-\frac{3}{64}\ln 7 - \frac{5}{65}\right) - \left(-\frac{3}{64}\ln\left(\frac{5}{3}\right) - \frac{1}{8}\right)$$

$$= \frac{1}{28} - \frac{3}{64}\ln\left(\frac{21}{5}\right).$$

$$\boxed{I_{10} = \int \frac{\ln\left(1 + \ln x\right)}{x}dx}$$

**Solution.**

Let us use the substitution $\ln x = t$, so that $dx/x = dt$. This gives the following

$$I_{10} = \int \ln(1 + t)dt.$$

For the integral, by using the integration by parts, let us assume

$$u = \ln(1 + t) \Longrightarrow du = \frac{dt}{1 + t}, \quad dv = dt \Longrightarrow v = t.$$

Thus, we have

$$I_{10} = t\ln(1 + t) - \int \frac{t}{1 + t}dt$$

$$= t\ln(1 + t) - \int \left(1 - \frac{1}{1 + t}\right)dt$$

$$= t\ln(1 + t) - t + \ln(1 + t) + C$$

$$= \ln(1 + t) + t\left(\ln(1 + t) - 1\right) + C$$

$$= \ln\left(1 + \ln x\right) + \ln x\left(\ln\left(1 + \ln x\right) - 1\right) + C,$$

where $C \in \mathbb{R}$.

$$\boxed{I_{11} = \int \sqrt{\csc x - \sin x}dx}$$

**Solution.**

For the given integral we can write the following

$$I_{11} = \int \sqrt{\frac{1}{\sin x} - \sin x}dx = \int \sqrt{\frac{1 - \sin^2 x}{\sin x}}dx = \int \frac{\sqrt{\cos^2 x}}{\sin x}dx$$

$$= \int \frac{|\cos^2 x|}{\sin x}dx = \int \sin^{-1/2} x \cos x dx = 2\sqrt{\sin x} + C,$$

where $C \in \mathbb{R}$.

Note that, in our solution, we considered $\cos x > 0$ (Be attentive, when the integral is definite!).

$$I_{12} = \int \frac{dx}{\sqrt{x^2 + 25}}$$

**Solution.**

By the fundamental rules of integration, We can simply calculate this integral, where

$$\int \frac{dx}{\sqrt{x^2 + 25}} = \ln \left| x + \sqrt{x^2 + 25} \right| + C,$$

where $C \in \mathbb{R}$.

$$I_{13} = \int_2^e \frac{\ln^2 x - 1}{x \ln^2 x} dx$$

**Solution.**

For the given integral we can write the following

$$
\begin{aligned}
I_{13} &= \int_2^e \left( \frac{\ln^2 x}{x \ln^2 x} - \frac{1}{x \ln^2 x} \right) dx = \int_2^e \left( \frac{1}{x} - \frac{1}{x} (\ln x)^{-2} \right) dx \\
&= \int_2^e \left( \frac{1}{x} - (\ln x)' (\ln x)^{-2} \right) dx = \left[ \ln x + \frac{1}{\ln x} \right]_2^e \\
&= 2 - \left( \ln 2 - \frac{1}{\ln 2} \right) = -\frac{(\ln 2 - 1)^2}{\ln 2}.
\end{aligned}
$$

$$I_{14} = \int e^{3x} \arctan (e^x) \, dx$$

**Solution.**

For the given integral we can write the following

$$I_{14} = \int e^{2x} \arctan (e^x) \, e^x dx.$$

Now, let us use the substitution $e^x = t$, so that $e^x dx = dt$. This gives the following

$$I_{14} = \int t^2 \arctan t \, dt.$$

Now, by using the integration by parts, let us assume

$$u = \arctan t \implies du = \frac{dt}{1 + t^2}, \quad dv = t^2 dt \implies v = \frac{t^3}{3}.$$

Thus, we have

$$I_{14} = \frac{t^3}{3}\arctan t - \frac{1}{3}\int \frac{t^3}{1+t^2}\,dt = \frac{t^3}{3}\arctan t - \frac{1}{3}\int\left(t - \frac{t}{1+t^2}\right)dt$$

$$= \frac{t^3}{3}\arctan t - \frac{1}{3}\left(\frac{t^2}{2} - \frac{1}{2}\ln\left(1+t^2\right)\right) + C$$

$$= \frac{t^3}{3}\arctan t - \frac{t^2}{6} + \frac{1}{6}\ln\left(1+t^2\right) + C$$

$$= \frac{e^{3x}}{3}\arctan\left(e^x\right) - \frac{e^{2x}}{6} + \frac{1}{6}\ln\left(1+e^{2x}\right) + C,$$

where $C \in \mathbb{R}$.

$$I_{15} = \int_0^4 \frac{|x-1|}{|x-2|+|x-3|}\,dx$$

**Solution.**

Let $f(x) := \frac{|x-1|}{|x-2|+|x-3|}$. Then, we have

$$f(x) = \frac{-(x-1)}{-(x-2)-(x-3)} = \frac{x-1}{2x-5}, \quad \forall x \in [0,1],$$

$$f(x) = \frac{x-1}{-(x-2)-(x-3)} = \frac{x-1}{-2x+5}, \quad \forall x \in [1,2],$$

$$f(x) = \frac{x-1}{(x-2)-(x-3)} = x-1, \quad \forall x \in [2,3],$$

$$f(x) = \frac{x-1}{(x-2)+(x-3)} = \frac{x-1}{2x-5}, \quad \forall x \in [3,4].$$

Therefore, for the given integral, we get

$$I_{15} = \int_0^1 \frac{x-1}{2x-5}\,dx - \int_1^2 \frac{x-1}{2x-5}\,dx + \int_2^3 (x-1)dx + \int_3^4 \frac{x-1}{2x-5}\,dx$$

$$= \int_0^1 \left(\frac{1}{2} + \frac{3}{2(2x-5)}\right)dx - \int_1^2 \left(\frac{1}{2} + \frac{3}{2(2x-5)}\right)dx$$

$$+ \int_2^3 (x-1)dx + \int_3^4 \left(\frac{1}{2} + \frac{3}{2(2x-5)}\right)dx$$

$$= \left[\frac{x}{2} + \frac{3}{4}\ln|2x-5|\right]_0^1 - \left[\frac{x}{2} + \frac{3}{4}\ln|2x-5|\right]_1^2 + \left[\frac{x^2}{2} - x\right]_2^3$$

$$+ \left[\frac{x}{2} + \frac{3}{4}\ln|2x-5|\right]_3^4$$

$$= \left( \frac{1}{2} + \frac{3}{4} \ln 3 - \frac{3}{4} \ln 5 \right) - \left( 1 - \frac{1}{2} - \frac{3}{4} \ln 3 \right) + \left( \frac{9}{2} - 3 - 2 \right)$$

$$+ \left( 2 + \frac{3}{4} \ln 3 - \frac{3}{2} \right) = 2 + \frac{9}{4} \ln 3 - \frac{3}{4} \ln 5.$$

$$I_{16} = \int_0^{2\pi} \frac{dx}{\sin^4 x + \cos^4 x}$$

**Solution.**

Let us calculate the indefinite integral

$$I := \int \frac{dx}{\sin^4 x + \cos^4 x}.$$

We have

$$\sin^2 x + \cos^2 x = 1,$$
$$\implies \sin^4 x + \cos^4 x + 2 \sin^2 x \cos^2 x = 1,$$
$$\implies \sin^4 x + \cos^4 x = 1 - 2 \sin^2 x \cos^2 x = 1 - 2 \left( \frac{1}{2} \sin(2x) \right)^2$$
$$= 1 - \frac{1}{2} \sin^2(2x).$$

Thus, we get

$$I = \int \frac{dx}{1 - \frac{1}{2} \sin^2(2x)} = 2 \int \frac{dx}{2 - \sin^2(2x)}.$$

For the last integral, let us use the substitution $\tan x = t$, so that $dx = \frac{dt}{1+t^2}$, and $\sin(2x) = \frac{2t}{1+t^2}$. This gives the following

$$I = 2 \int \frac{dt}{(1+t^2)\left(2 - \frac{4t^2}{(1+t^2)^2}\right)} = 2 \int \frac{1+t^2}{2(1+t^2)^2 - 4t^2} dt = 2 \int \frac{1+t^2}{2 + 2t^4} dt$$

$$= \int \frac{1+t^2}{1+t^4} dt = \int \frac{t^2 \left(1 + \frac{1}{t^2}\right)}{t^2 \left(t^2 + \frac{1}{t^2}\right)} dt = \int \frac{1 + \frac{1}{t^2}}{\left(t - \frac{1}{t}\right)^2 + 2} dt.$$

Now, for the last integral, let us use the substitution $t - \frac{1}{t} = u$, so that $\left(1 + \frac{1}{t^2}\right) dt = du$. This gives the following

$$I = \int \frac{du}{2 + u^2} = \frac{1}{\sqrt{2}} \arctan\left(\frac{u}{\sqrt{2}}\right) + C = \frac{1}{\sqrt{2}} \arctan\left(\frac{t^2 - 1}{\sqrt{2}t}\right) + C$$

$$= \frac{1}{\sqrt{2}} \arctan\left(\frac{\tan^2 x - 1}{\sqrt{2}\tan x}\right) + C,$$

where $C \in \mathbb{R}$.

We note that the function $f(x) := \arctan\left(\frac{\tan^2 x - 1}{\sqrt{2}\tan x}\right)$ is not continuous at the points $0, \frac{\pi}{2}, \pi, \frac{3\pi}{2}, 2\pi$ (see Fig. 8.1). Therefore, we can write

$$I_{16} = \int_0^{\frac{\pi}{2}} \frac{dx}{\sin^4 x + \cos^4 x} + \int_{\frac{\pi}{2}}^{\pi} \frac{dx}{\sin^4 x + \cos^4 x} + \int_{\pi}^{\frac{3\pi}{2}} \frac{dx}{\sin^4 x + \cos^4 x}$$

$$+ \int_{\frac{3\pi}{2}}^{2\pi} \frac{dx}{\sin^4 x + \cos^4 x}$$

$$= \frac{1}{\sqrt{2}} \lim_{\substack{a_1 \to 0^+ \\ b_1 \to (\pi/2)^-}} \left[\arctan\left(\frac{\tan^2 x - 1}{\sqrt{2}\tan x}\right)\right]_{a_1}^{b_1}$$

$$+ \frac{1}{\sqrt{2}} \lim_{\substack{a_2 \to (\pi/2)^+ \\ b_2 \to \pi^-}} \left[\arctan\left(\frac{\tan^2 x - 1}{\sqrt{2}\tan x}\right)\right]_{a_2}^{b_2}$$

$$+ \frac{1}{\sqrt{2}} \lim_{\substack{a_3 \to \pi^+ \\ b_3 \to (3\pi/2)^-}} \left[\arctan\left(\frac{\tan^2 x - 1}{\sqrt{2}\tan x}\right)\right]_{a_3}^{b_3}$$

$$+ \frac{1}{\sqrt{2}} \lim_{\substack{a_4 \to (3\pi/2)^+ \\ b_4 \to 2\pi^-}} \left[\arctan\left(\frac{\tan^2 x - 1}{\sqrt{2}\tan x}\right)\right]_{a_4}^{b_4}$$

$$= \frac{1}{\sqrt{2}}\left(\left(\frac{\pi}{2} + \frac{\pi}{2}\right) + \left(\frac{\pi}{2} + \frac{\pi}{2}\right) + \left(\frac{\pi}{2} + \frac{\pi}{2}\right) + \left(\frac{\pi}{2} + \frac{\pi}{2}\right)\right) = 2\sqrt{2}\pi.$$

Where we used the following limits (see Fig. 8.1)

$$\lim_{x \to 0^+} \frac{\tan^2 x - 1}{\sqrt{2}\tan x} = -\infty, \qquad \lim_{x \to (\pi/2)^-} \frac{\tan^2 x - 1}{\sqrt{2}\tan x} = +\infty,$$

$$\lim_{x \to (\pi/2)^+} \frac{\tan^2 x - 1}{\sqrt{2}\tan x} = -\infty, \qquad \lim_{x \to \pi^-} \frac{\tan^2 x - 1}{\sqrt{2}\tan x} = +\infty,$$

$$\lim_{x \to \pi^+} \frac{\tan^2 x - 1}{\sqrt{2}\tan x} = -\infty, \qquad \lim_{x \to (3\pi/2)^-} \frac{\tan^2 x - 1}{\sqrt{2}\tan x} = +\infty,$$

$$\lim_{x \to (3\pi/2)^+} \frac{\tan^2 x - 1}{\sqrt{2}\tan x} = -\infty, \qquad \lim_{x \to 2\pi^-} \frac{\tan^2 x - 1}{\sqrt{2}\tan x} = +\infty.$$

**Remark.** The integrand $f(x) := \frac{dx}{\sin^4 x + \cos^4 x}$ in integral $I_{16}$, is a periodic function with primitive period $P = \frac{\pi}{2}$ (see Fig. 8.2). Therefore, we can write

$$I_{16} = 4\int_0^{\pi/2} \frac{dx}{\sin^4 x + \cos^4 x}\, dx = \frac{4}{\sqrt{2}} \lim_{\substack{a_1 \to 0^+ \\ b_1 \to (\pi/2)^-}} \left[\arctan\left(\frac{\tan^2 x - 1}{\sqrt{2}\tan x}\right)\right]_{a_1}^{b_1}$$

$$= \frac{4}{\sqrt{2}}\left(\frac{\pi}{2} + \frac{\pi}{2}\right) = 2\sqrt{2}\pi.$$

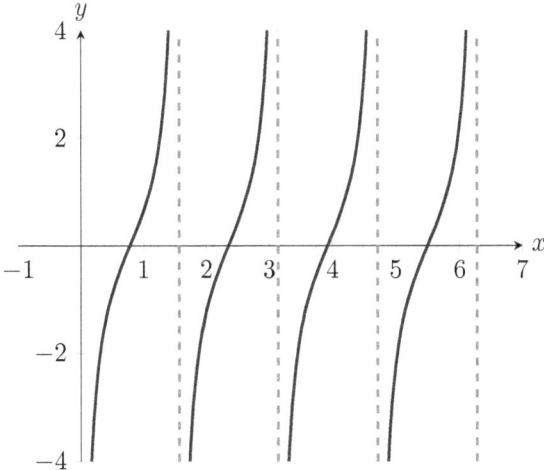

Figure 8.1: The graph of $g(x) := \frac{\tan^2 x - 1}{\sqrt{2}\tan x}$ in $]0, 2\pi[$.

$$I_{17} = \int \frac{1 + e^x}{1 - e^x}dx$$

## Solution.

For the given integral we can write the following

$$I_{17} = \int \frac{1 + e^x}{1 - e^x}dx = \int \frac{dx}{1 - e^x} + \int \frac{e^x}{1 - e^x}dx$$

$$= \int \frac{e^{-x}}{e^{-x} - 1}dx + \int \frac{e^x}{1 - e^x}dx$$

$$= -\int \frac{-e^{-x}}{e^{-x} - 1}dx - \int \frac{-e^x}{1 - e^x}dx$$

$$= -\ln\left|e^{-x} - 1\right| - \ln\left|1 - e^x\right| + C$$

$$= -\ln\left|\frac{1 - e^x}{e^x}\right| - \ln\left|1 - e^x\right| + C$$

$$= -\ln\left|1 - e^x\right| + \ln\left(e^x\right) - \ln\left|1 - e^x\right| + C$$

$$= x - 2\ln\left|1 - e^x\right| + C,$$

where $C \in \mathbb{R}$.

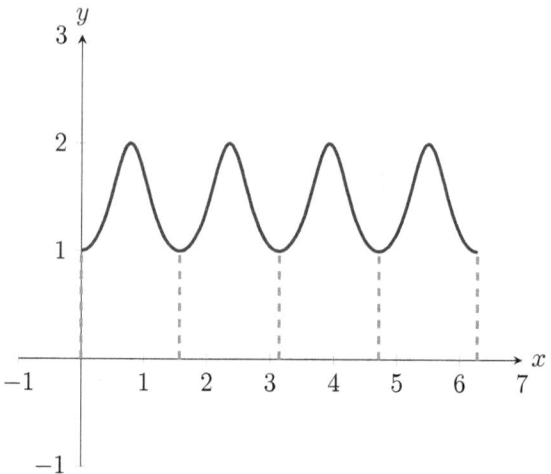

Figure 8.2:  The graph of $f(x) := \frac{1}{\sin^4 x + \cos^4 x}$ in $[0, 2\pi]$.

$$I_{18} = \int \tan^4 x \, dx$$

**Solution.**

For the given integral we can write the following

$$I_{18} = \int \tan^2 x \left(1 + \tan^2 x - 1\right) dx$$

$$= \int \left(\tan^2 x \left(1 + \tan^2 x\right) - \tan^2 x\right) dx$$

$$= \int \left(\tan^2 x \frac{1}{\cos^2 x} - \frac{1 - \cos^2 x}{\cos^2 x}\right) dx$$

$$= \int \left(\tan^2 x \frac{1}{\cos^2 x} - \frac{1}{\cos^2 x} + 1\right) dx$$

$$= \int \left(\tan^2 x (\tan x)' - (\tan x)' + 1\right) dx$$

$$= \frac{1}{3} \tan^3 x - \tan x + x + C,$$

where $C \in \mathbb{R}$.

$$I_{19} = \int \sin x \tan^2 x \, dx$$

**Solution.**

For the given integral we can write the following

$$I_{19} = \int \sin x \frac{\sin^2 x}{\cos^2 x} dx = \int \left( \frac{1 - \cos^2 x}{\cos^2 x} \right) \sin x dx$$

$$= \int \left( \frac{1}{\cos^2 x} - 1 \right) \sin x dx.$$

For the last integral, let us use the substitution $\cos x = t$, so that $-\sin x dx = dt$. This gives the following

$$I_{19} = -\int \left( \frac{1}{t^2} - 1 \right) dt = \frac{1}{t} + t + C = \frac{1}{\cos x} + \cos x + C,$$

where $C \in \mathbb{R}$.

$$\boxed{I_{20} = \int \frac{x + 1}{x^2 + 2x + 3} dx}$$

**Solution.**

For the given integral we can write the following

$$I_{20} = \frac{1}{2} \int \frac{2x + 2}{x^2 + 2x + 3} dx = \frac{1}{2} \ln \left| x^2 + 2x + 3 \right| + C,$$

where $C \in \mathbb{R}$.

# Chapter 9

# The Solutions to the 2016 MIT Integration Bee, Qualifying Test

$$I_1 = \int \tanh x dx$$

**Solution.**

For the given integral we can write the following

$$I_1 = \int \tanh x dx = \int \frac{\sinh x}{\cosh x} dx = \ln |\cosh x| + C,$$

where $C \in \mathbb{R}$.

$$I_2 = \int_{-4}^{4} |x^3 - x| \, dx$$

**Solution.**

We note that the integrand $f(x) := |x^3 - x|$ is an even function in $[-4, 4]$ (see Fig. 9.1). Therefore, we can write

$$I_2 = 2 \int_0^4 |x^3 - x| \, dx = 2 \left( \int_0^1 (x - x^3) \, dx + \int_1^4 (x^3 - x) \, dx \right)$$

$$= 2 \left( \left[ \frac{x^2}{2} - \frac{x^4}{4} \right]_0^1 + \left[ \frac{x^4}{4} - \frac{x^2}{2} \right]_1^4 \right) = 2 \left( \left( \frac{1}{2} - \frac{1}{4} \right) + \left( 64 - 8 - \frac{1}{4} + \frac{1}{2} \right) \right)$$

$$= 113.$$

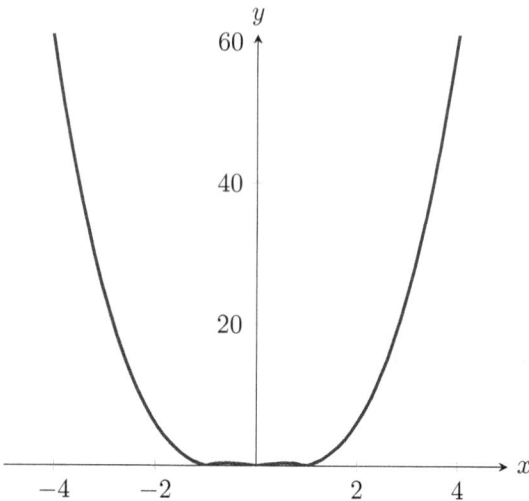

Figure 9.1:  The graph of $f(x) = |x^3 - x|$ in $[-4, 4]$.

$$I_3 = \int_1^e \ln\left(\sqrt{x}\right) dx$$

**Solution.**
Let us use the substitution $\sqrt{x} = t$, so that $dx = 2\sqrt{x}\,dt = 2t\,dt$. If $x = 1$, then $t = 1$, and if $x = e$, then $t = \sqrt{e}$. With these new values, we get the following

$$I_3 = 2\int_1^{\sqrt{e}} t\ln t\,dt.$$

Now, by using the integration by parts, let us assume

$$u = \ln t \Longrightarrow du = dt/t, \quad dv = t\,dt \Longrightarrow v = t^2/2.$$

Thus, we have

$$I_3 = 2\left(\left[\frac{t^2}{2}\ln t\right]_1^{\sqrt{e}} - \frac{1}{2}\int_1^{\sqrt{e}} t\,dt\right) = 2\left(\frac{e}{4} - \frac{1}{2}\left[\frac{t^2}{2}\right]_1^{\sqrt{e}}\right)$$
$$= 2\left(\frac{e}{4} - \frac{1}{2}\left(\frac{e}{2} - \frac{1}{2}\right)\right) = \frac{1}{2}.$$

$$I_4 = \int\left(e^{e^x + e^{-x} + x} - e^{e^x + e^{-x} - x}\right) dx$$

**Solution.**

For the given integral we can write the following

$$I_4 = \int \left( e^{e^x + e^{-x}} e^x - e^{e^x + e^{-x}} e^{-x} \right) dx = \int e^{e^x + e^{-x}} \left( e^x - e^{-x} \right) dx.$$

For the last integral, let us use the substitution $e^x - e^{-x} = t$, so that $(e^x - e^{-x}) dx = dt$. This gives the following

$$I_4 = \int e^t dt = e^t + C = e^{e^x + e^{-x}} + C,$$

where $C \in \mathbb{R}$.

$$I_5 = \int \frac{\ln(\ln x)}{x \ln x} dx$$

**Solution.**

Let us use the substitution $\ln x = t$, so that $dx/x = dt$. This gives the following

$$I_5 = \int \frac{\ln t}{t} dt = \int (\ln t)' \ln t \, dt = \frac{1}{2} \ln^2 t + C = \frac{1}{2} (\ln(\ln x))^2 + C,$$

where $C \in \mathbb{R}$.

$$I_6 = \int_0^{\pi/3} \frac{dx}{1 + \tan^2 x}$$

**Solution.**

For the given integral we can write the following

$$I_6 = \int_0^{\pi/3} \frac{dx}{1 + \frac{\sin^2 x}{\cos^2 x}} = \int_0^{\pi/3} \frac{\cos^2 x}{\sin^2 x + \cos^2 x} dx = \int_0^{\pi/3} \cos^2 x \, dx$$

$$= \frac{1}{2} \int_0^{\pi/3} (1 + \cos(2x)) \, dx = \frac{1}{2} \left[ x + \frac{1}{2} \sin(2x) \right]_0^{\pi/3}$$

$$= \frac{1}{2} \left( \frac{\pi}{3} + \frac{1}{2} \sin\left( \frac{2\pi}{3} \right) \right) = \frac{\pi}{6} + \frac{\sqrt{3}}{8}.$$

$$I_7 = \int_{-27}^{27} \arcsin\left( \sqrt[3]{x}/3 \right) dx$$

**Solution.**

The function $f(x) := \arcsin\left( \sqrt[3]{x}/3 \right)$ is an odd in $[-27, 27]$. Therefore $I_7 = 0$.

$$I_8 = \int_{50}^{100} \lfloor \log_2 x \rfloor \, dx$$

**Solution.**

Let us find the values of $x$ in the interval $[50, 100]$, for which $\log_2 x \in \mathbb{N} = \{1, 2, \ldots\}$. For this, we have the following equation by $x$

$$\log_2 x = n; \quad n = 1, 2, \ldots \quad \Longrightarrow \quad x = 2^n; \quad n = 1, 2, \ldots$$

Therefore, when $x \in \{2^n; n = 1, 2, \ldots\}$, we have $\log_2 x \in \mathbb{N}$. We note that $2^5 = 32 \notin [50, 100]$, $2^6 = 64 \in [50, 100]$ and $2^7 = 128 \notin [50, 100]$. Thus, we find the following

$$\lfloor \log_2 x \rfloor = 5; \quad \forall x \in [50, 64[, \quad \text{and} \quad \lfloor \log_2 x \rfloor = 6; \quad \forall x \in [64, 100[.$$

Therefore, for the given integral, we get

$$I_8 = \int_{50}^{100} \lfloor \log_2 x \rfloor \, dx = \int_{50}^{64} 5dx + \int_{64}^{100} 6dx$$
$$= 5(64 - 50) + 6(100 - 64) = 288.$$

$$I_9 = \int (e^x \cos x - e^x \sin x) \, dx$$

**Solution 1.**

Note that

$$(e^x \cos x)' = e^x \cos x - e^x \sin x.$$

Thus, we have

$$I_9 = \int (e^x \cos x - e^x \sin x) \, dx = \int (e^x \cos x)' \, dx = e^x \cos x + C,$$

where $C \in \mathbb{R}$.

**Solution 2.**

For the given integral we can write the following

$$I_9 = \underbrace{\int e^x \cos x dx}_{:=J} - \int e^x \sin x dx.$$

For the integral $J$, by using the integration by parts, let us assume

$$u = \cos x \Longrightarrow du = -\sin x dx, \quad dv = e^x dx \Longrightarrow v = e^x.$$

Thus, we have

$$I_9 = e^x \cos x + \int e^x \sin x dx - \int e^x \sin x dx = e^x \cos x + C,$$

where $C \in \mathbb{R}$.

$$\boxed{I_{10} = \int_0^\infty x^3 e^{-x^2} dx}$$

**Solution 1.**

For the given integral we can write the following

$$I_{10} = \int_0^\infty x^3 e^{-x^2} dx = \lim_{a \to \infty} \int_0^a x^3 e^{-x^2} dx.$$

Let us calculate the indefinite integral

$$I := \int x^3 e^{-x^2} dx = \int x^2 e^{-x^2} x dx.$$

For this, let us use the substitution $x^2 = t$, so that $2x dx = dt$, and then let us use the integration by parts (one time). we get the following

$$I = \frac{1}{2} \int t e^{-t} = \frac{1}{2} \left( -t e^{-t} - e^{-t} \right) + C = -\frac{1}{2} e^{-t} (t+1) + C = -\frac{1}{2} e^{-x^2} (1 + x^2) + C,$$

where $C \in \mathbb{R}$.

Therefore, for the given integral $I_{10}$, we have

$$I_{10} = -\frac{1}{2} \lim_{a \to \infty} \left[ e^{-x^2} (1 + x^2) \right]_0^a = -\frac{1}{2} \lim_{a \to \infty} \left( \frac{a^2 + 1}{e^{a^2}} - 1 \right) = \frac{1}{2}.$$

**Solution 2.**

Remember that the gamma function has the following form (see Subsec. 1.8.1)

$$\Gamma(a) = \int_0^\infty x^{a-1} e^{-x} dx, \quad \forall a > 0.$$

Therefore, for the given integral $I_{10}$, let us use the substitution $x^2 = t$, so that $x dx = dt/2$. If $x = 0$ then $t = 0$ and if $x \to \infty$ then $t \to \infty$. This gives the following

$$I_{10} = \frac{1}{2} \int_0^\infty t e^{-t} dt = \frac{1}{2} \int_0^\infty t^{2-1} e^{-t} dt = \frac{1}{2} \Gamma(2) = \frac{1}{2} (1!) = \frac{1}{2}.$$

$$I_{11} = \int \left( \left( 2xe^{x^2} + 1 \right) \cos x - \left( e^{x^2} + x \right) \sin x \right) dx$$

**Solution 1.**

Note that

$$\left( \left( e^{x^2} + x \right) \cos x \right)' = \left( \left( 2xe^{x^2} + 1 \right) \cos x - \left( e^{x^2} + x \right) \sin x \right).$$

Therefore, we get

$$I_{11} = \int \left( \left( e^{x^2} + x \right) \cos x \right)' dx = \left( e^{x^2} + x \right) \cos x + C,$$

where $C \in \mathbb{R}$.

**Solution 2.**

For the given integral we can write the following

$$I_{11} = 2 \int xe^{x^2} \cos x dx + \int \cos x dx - \underbrace{\int e^{x^2} \sin x dx}_{:=J_1} - \underbrace{\int x \sin x dx}_{:=J_2}$$

For the integral $J_1$, by using the integration by parts, let us assume

$$u = e^{x^2} \implies du = 2xe^{x^2} dx, \quad dv = \sin x dx \implies v = -\cos x.$$

Thus, we have

$$J_1 = -e^{x^2} \cos x + 2 \int x \cos x e^{x^2} dx.$$

For the integral $J_2$, by using the integration by parts ($u = x, dv = \sin x dx$), we find

$$J_2 = -x \cos x + \sin x + C_1,$$

where $C_1 \in \mathbb{R}$.

Therefore for the given integral $I_{11}$, we get

$$I_{11} = 2 \int xe^{x^2} \cos x dx + \sin x - \left( -e^{x^2} \cos x + 2 \int xe^{x^2} \cos x dx \right)$$
$$- (-x \cos x + \sin x + C_1)$$
$$= 2 \int xe^{x^2} \cos x dx + \sin x + e^{x^2} \cos x - 2 \int xe^{x^2} \cos x dx$$
$$+ x \cos x - \sin x - C_1$$
$$= x \cos x + e^{x^2} \cos x + C = \left( x + e^{x^2} \right) \cos x + C,$$

where $C = -C_1 \in \mathbb{R}$.

$$I_{12} = \int \left(1 + x^{1/2} + x^{1/3}\right)\left(1 + x^{-1/2} + x^{-1/3}\right) dx$$

**Solution.**

For the given integral we can write the following

$$I_{12} = \int \left(3 + x^{-1/2} + x^{-1/3} + x^{1/2} + x^{1/3} + x^{1/6} + x^{-1/6}\right) dx$$

$$= 3x + 2x^{1/2} + \frac{3}{2}x^{2/3} + \frac{2}{3}x^{3/2} + \frac{3}{4}x^{4/3} + \frac{6}{7}x^{7/6} + \frac{6}{5}x^{5/6} + C,$$

where $C \in \mathbb{R}$.

$$I_{13} = \int \sin\left(\sin x\right) \cos\left(\sin x\right) \cos x \, dx$$

**Solution.**

Let us use the substitution $\sin x = t$, so that $\cos x \, dx = dt$. This gives the following

$$I_{13} = \int \sin t \cos t \, dt = \frac{1}{2} \int \sin(2t) dt = -\frac{1}{4} \cos(2t) + C_1, \quad \text{(where } C_1 \in \mathbb{R})$$

$$= -\frac{1}{4}\cos\left(2\sin x\right) + C_1 = -\frac{1}{4}\left(\cos^2\left(\sin x\right) - \sin^2\left(\sin x\right)\right) + C_1$$

$$= -\frac{1}{4}\left(1 - 2\sin^2\left(\sin x\right)\right) + C_1 = -\frac{1}{4} + \frac{1}{2}\sin^2\left(\sin x\right) + C_1$$

$$= \frac{1}{2}\sin^2\left(\sin x\right) + C,$$

where $C = C_1 - \frac{1}{4} \in \mathbb{R}$.

$$I_{14} = \int \left(\frac{\cos x + \sin x}{x^2} + \frac{\sin x - \cos x}{x}\right) dx$$

**Solution 1.**

Note that

$$\left(\frac{1}{x}(\sin x + \cos x)\right)' = -\frac{\sin x + \cos x}{x^2} + \frac{\cos x - \sin x}{x}$$

$$= -\left(\frac{\sin x - \cos x}{x} + \frac{\sin x + \cos x}{x^2}\right).$$

Thus, we have

$$
I_{14} = - \int \left( \frac{\sin x + \cos x}{x} \right)' dx = - \frac{1}{x} (\sin x + \cos x) + C,
$$

where $C \in \mathbb{R}$.

**Solution 2.**

For the given integral we can write the following

$$
I_{14} = \int \frac{\cos x + \sin x}{x^2} dx + \underbrace{\int \frac{\sin x - \cos x}{x} dx}_{:=J} .
$$

For the integral $J$, by using the integration by parts, let us assume

$$
u = \frac{1}{x} \implies du = -\frac{dx}{x^2}, \quad dv = (\sin x - \cos x)dx \implies v = -\cos x - \sin x.
$$

Thus, we have

$$
J = -\frac{1}{x} (\sin x + \cos x) - \int \frac{\sin x + \cos x}{x^2} dx.
$$

Therefore, for the given integral $I_{14}$, we get

$$
\begin{aligned}
I_{14} &= \int \frac{\sin x + \cos x}{x^2} dx - \frac{1}{x} (\sin x + \cos x) - \int \frac{\sin x + \cos x}{x^2} dx \\
&= -\frac{1}{x} (\sin x + \cos x) + C,
\end{aligned}
$$

where $C \in \mathbb{R}$.

$$
\boxed{I_{15} = \int x^3 \sqrt{x^2 + 1} dx}
$$

**Solution.**

For the given integral we can write the following

$$
I_{15} = \int x^3 \sqrt{x^2 + 1} dx = \int x^2 \sqrt{x^2 + 1} x dx.
$$

For the last integral, let us use the substitution $x^2 = t$, so that $2x dx = dt$. This gives the following

$$I_{15} = \frac{1}{2} \int t\sqrt{1+t}\,dt = \frac{1}{2} \int (t+1-1)\sqrt{1+t}\,dt$$

$$= \frac{1}{2} \int \left( (1+t)^{3/2} - (1+t)^{1/2} \right) dt$$

$$= \frac{1}{2} \left( \frac{2}{5}(1+t)^{5/2} - \frac{2}{3}(1+t)^{3/2} \right) + C$$

$$= \frac{1}{5}(1+t)^{5/2} - \frac{1}{3}(1+t)^{3/2} + C$$

$$= \frac{1}{15}(1+t)^{3/2}\left( 3(1+t) - 5 \right) + C$$

$$= \frac{1}{15}(1+x^2)^{3/2}(3x^2 - 2) + C,$$

where $C \in \mathbb{R}$.

**Remark.** For the integral

$$\int t\sqrt{1+t}\,dt,$$

we can use the integration by parts. We assume

$$u = t \Longrightarrow du = dt, \quad dv = \sqrt{1+t}\,dt \Longrightarrow v = \frac{2}{3}(1+t)^{3/2}.$$

Thus, we get

$$I_{15} = \frac{1}{2} \left( \frac{2}{3}t(1+t)^{3/2} - \frac{2}{3}\int (1+t)^{3/2} dt \right)$$

$$= \frac{1}{2} \left( \frac{2}{3}t(1+t)^{3/2} - \frac{2}{3}\frac{2}{5}(1+t)^{5/2} \right) + C$$

$$= \frac{1}{3}t(1+t)^{3/2} - \frac{2}{15}(1+t)^{5/2} + C$$

$$= \frac{1}{3}x^2(1+x^2)^{3/2} - \frac{2}{15}(1+x^2)^{5/2} + C,$$

where $C \in \mathbb{R}$.

$$\boxed{I_{16} = \int \frac{x}{x^4 + x^2 + 1}\,dx}$$

**Solution.**

For the given integral we can write the following

$$I_{16} = \int \frac{x}{x^4 + x^2 + \frac{1}{4} + \frac{3}{4}}\,dx = \int \frac{x}{\left(x^2 + \frac{1}{2}\right)^2 + \frac{3}{4}}\,dx.$$

For the last integral, let us use the substitution $x^2 + \frac{1}{2} = t$, so that $2x\,dx = dt$. This gives the following

$$I_{16} = \frac{1}{2} \int \frac{dt}{t^2 + \left(\sqrt{3}/2\right)^2} = \frac{1}{\sqrt{3}} \arctan\left(\frac{2t}{\sqrt{3}}\right) + C$$
$$= \frac{1}{\sqrt{3}} \arctan\left(\frac{2x^2 + 1}{\sqrt{3}}\right) + C,$$

where $C \in \mathbb{R}$.

$$\boxed{I_{17} = \int e^{e^{2016x} + 6048x}\,dx}$$

**Solution.**

For the given integral we can write the following

$$I_{17} = \int e^{e^{2016x}} e^{3(2016)x}\,dx.$$

Let us use the substitution $e^{2016x} = t$, so that

$$2016e^{2016x}\,dx = dt \implies dx = \frac{dt}{2016e^{2016x}} = \frac{dt}{2016t}.$$

This gives the following

$$I_{17} = \int \frac{t^3 e^t}{2016t}\,dt = \frac{1}{2016} \int t^2 e^t dt.$$

By using the integration by parts two times (at the first time, we assume $u = t^2$, $dv = e^t dt$), we get the following

$$I_{17} = \frac{1}{2016}(t^2 e^t - 2te^t + 2e^t) + C = \frac{e^t}{2016}\left(t^2 - 2t + 2\right) + C$$
$$= \frac{e^{2016x}}{2016}\left(e^{4032x} - 2e^{2016x} + 2\right) + C,$$

where $C \in \mathbb{R}$.

$$\boxed{I_{18} = \int_{\pi/3}^{\pi/2} \frac{1 - \cos x}{\sin x}\,dx}$$

**Solution 1.**

For the given integral we can write the following

$$I_{18} = \int_{\pi/3}^{\pi/2} \left( \frac{1}{\sin x} - \frac{\cos x}{\sin x} \right) dx = [\ln |\tan (x/2)| - \ln | \sin x|]_{\pi/3}^{\pi/2}$$
$$= (\ln (\tan (\pi/4)) - \ln (\sin (\pi/2))) - (\ln (\tan (\pi/6)) - \ln (\tan (\pi/3)))$$
$$= - \left( \ln \left( 1/\sqrt{3} \right) - \ln \left( \sqrt{3}/2 \right) \right) = 2 \ln \sqrt{3} - \ln 2 = \ln 3 - \ln 2 = \ln (3/2).$$

**Solution 2.**

For the given integral we can write the following

$$I_{18} = \int_{\pi/3}^{\pi/2} \frac{2 \sin^2 (x/2)}{2 \sin (x/2) \cos (x/2)} dx = \int_{\pi/3}^{\pi/2} \frac{\sin (x/2)}{\cos (x/2)} dx$$
$$= -2 \int_{\pi/3}^{\pi/2} \frac{(-1/2) \sin (x/2)}{\cos (x/2)} dx = -2 \left[ \ln \left| \cos \left( \frac{x}{2} \right) \right| \right]_{\pi/3}^{\pi/2}$$
$$= -2 (\ln (\cos(\pi/4)) - \ln (\cos(\pi/6)))$$
$$= -2 \left( \ln \left( 1/\sqrt{2} \right) - \ln \left( \sqrt{3}/2 \right) \right) = -2 \left( \frac{\ln 2}{2} - \frac{\ln 3}{2} \right) = \ln (3/2).$$

$$\boxed{I_{19} = \int \frac{dx}{1 - x + x^2 - x^3}}$$

**Solution.**

For the given integral we can write the following

$$I_{19} = \int \frac{dx}{1 - x + x^2 - x^3} = - \int \frac{dx}{x^3 - x^2 + x - 1}.$$

According to the method of partial fractions, we have

$$f(x) := \frac{1}{x^3 - x^2 + x - 1} = \frac{1}{(x - 1)(x^2 + 1)} = \frac{A}{x - 1} + \frac{Bx + C}{x^2 + 1}.$$

Thus,

$$1 = A(x^2 + 1) + (Bx + C)(x - 1) = (A + B)x^2 + (C - B)x + A - C. \quad (9.1)$$

By the corresponding between the sides of the equality (9.1), we find the following linear system of equations

$$\begin{cases} A + B = 0, \\ C - B = 0, \\ A - C = 1. \end{cases}$$

By solving this system, we find $A = 1/2$, $B = -1/2$, $C = -1/2$. Therefore, for the given integral, we find

$$
\begin{aligned}
I_{19} &= -\int \left( \frac{1}{2(x-1)} - \frac{1}{2}\left(\frac{x+1}{x^2+1}\right) \right) dx \\
&= \int \left( -\frac{1}{2}\frac{1}{x-1} + \frac{1}{4}\frac{2x}{x^2+1} + \frac{1}{2}\frac{1}{x^2+1} \right) dx \\
&= -\frac{1}{2}\ln|x-1| + \frac{1}{4}\ln(1+x^2) + \frac{1}{2}\arctan x + C' \\
&= \frac{1}{2}\left( \arctan x - \ln|1-x| + \ln\left(\sqrt{1+x^2}\right) \right) + C',
\end{aligned}
$$

where $C' \in \mathbb{R}$.

$$
\boxed{I_{20} = \int_0^\infty \frac{dx}{2 + \cosh x}}
$$

**Solution 1.**

For the given integral we can write the following

$$
\begin{aligned}
I_{20} &= \int_0^\infty \frac{dx}{2 + \frac{e^x + e^{-x}}{2}} = 2\int_0^\infty \frac{dx}{e^x + e^{-x} + 4} \\
&= 2\int_0^\infty \frac{e^x}{\left(e^x\right)^2 + 4e^x + 1} dx \\
&= \lim_{a\to\infty} \int_0^a \frac{e^x}{\left(e^x\right)^2 + 4e^x + 1} dx.
\end{aligned}
$$

Now, let us calculate the indefinite integral

$$
I := \int \frac{e^x}{\left(e^x\right)^2 + 4e^x + 1} dx.
$$

For this, let us use the substitution $e^x = t$, so that $e^x dx = dt$. This gives the following

$$
\begin{aligned}
I &= \int \frac{dt}{t^2 + 4t + 1} = \int \frac{dt}{(t+2)^2 - 3} \\
&= -\int \frac{dt}{\left(\sqrt{3}\right)^2 - (t+2)^2} = -\frac{1}{2\sqrt{3}}\ln\left|\frac{\sqrt{3}+t+2}{\sqrt{3}-t-2}\right| + C \\
&= -\frac{1}{2\sqrt{3}}\ln\left|\frac{\sqrt{3}+2+e^x}{\sqrt{3}-2-e^x}\right| + C,
\end{aligned}
$$

where $C \in \mathbb{R}$.

Therefore, for the given integral $I_{20}$, we get

$$I_{20} = -\frac{2}{2\sqrt{3}} \lim_{a \to \infty} \left( \left[ \ln \left| \frac{\sqrt{3} + 2 + e^x}{\sqrt{3} - 2 - e^x} \right| \right]_0^a \right)$$

$$= -\frac{1}{\sqrt{3}} \lim_{a \to \infty} \left( \ln \left| \frac{\sqrt{3} + 2 + e^a}{\sqrt{3} - 2 - e^a} \right| - \ln \left| \frac{\sqrt{3} + 3}{\sqrt{3} - 3} \right| \right)$$

$$= \frac{1}{\sqrt{3}} \ln \left| \frac{\sqrt{3} + 3}{\sqrt{3} - 3} \right| = \frac{1}{\sqrt{3}} \ln \left( \frac{(3 + \sqrt{3})^2}{6} \right)$$

$$= \frac{1}{\sqrt{3}} \ln \left( 2 + \sqrt{3} \right).$$

**Solution 2.**

Let us use the substitution $\tanh \left( \frac{x}{2} \right) = t$, so that $dx = \frac{2dt}{1-t^2}$ and $\cosh x = \frac{1+t^2}{1-t^2}$. Remember that $\tanh \left( \frac{x}{2} \right) = \frac{e^{x/2} - e^{-x/2}}{e^{x/2} + e^{-x/2}}$, thus if $x = 0$, then $t = 0$, and if $x \to \infty$, then $t \to 1$. Therefore, we get

$$I_{20} = \int_0^1 \frac{\frac{2dt}{1-t^2}}{2 + \frac{1+t^2}{1-t^2}} = 2 \int_0^1 \frac{dt}{3 - t^2} = 2 \frac{1}{2\sqrt{3}} \left[ \ln \left| \frac{\sqrt{3} + t}{\sqrt{3} - t} \right| \right]_0^1$$

$$= \frac{1}{\sqrt{3}} \ln \left( \frac{\sqrt{3} + 1}{\sqrt{3} - 1} \right) = \frac{1}{\sqrt{3}} \ln \left( \frac{(\sqrt{3} + 1)^2}{2} \right)$$

$$= \frac{1}{\sqrt{3}} \ln \left( 2 + \sqrt{3} \right).$$

# Chapter 10

# The Solutions to the 2017 MIT Integration Bee, Qualifying Test

$$I_1 = \int \frac{x^2}{\sqrt{x^3 + 2}} dx$$

**Solution.**

For the given integral we can write the following

$$I_1 = \int \frac{x^2}{\sqrt{x^3 + 2}} dx = \frac{1}{3} \int \frac{3x^2}{\sqrt{x^3 + 2}} dx = \frac{2}{3} \sqrt{x^3 + 2} + C,$$

where $C \in \mathbb{R}$.

$$I_2 = \int_1^\infty \frac{\ln x}{x^2} dx$$

**Solution.**

For the given integral we can write the following

$$I_2 = \int_1^\infty \frac{\ln x}{x^2} dx = \lim_{a \to \infty} \int_1^a \frac{\ln x}{x^2} dx.$$

By using the integration by parts, let us assume

$$u = \ln x \implies du = \frac{dx}{x}, \quad dv = \frac{dx}{x^2} \implies v = -\frac{1}{x}.$$

Thus, we have

$$I_2 = \lim_{a \to \infty} \left( \left[ -\frac{\ln x}{x} \right]_1^a + \int_1^a \frac{dx}{x^2} \right) = \lim_{a \to \infty} \left( -\frac{\ln a}{a} - \left[ \frac{1}{x} \right]_1^a \right)$$

$$= \lim_{a \to \infty} \left( -\frac{\ln a}{a} - \frac{1}{a} + 1 \right) = 1.$$

$$\boxed{I_3 = \int \operatorname{sech} x\, dx}$$

## Solution.

For the given integral we can write the following

$$I_3 = \int \frac{dx}{\cosh x} = 2 \int \frac{dx}{e^x + e^{-x}} = 2 \int \frac{e^x}{(e^x)^2 + 1}\, dx.$$

Let us use the substitution $e^x = t$, so that $e^x dx = dt$. This gives the following

$$I_3 = 2 \int \frac{dt}{1 + t^2} = 2\arctan t + C = 2\arctan(e^x) + C,$$

where $C \in \mathbb{R}$.

$$\boxed{I_4 = \int x^3 e^{x^2}\, dx}$$

## Solution.

For the given integral we can write the following

$$I_4 = \int x^3 e^{x^2}\, dx = \int x^2 e^{x^2} x\, dx.$$

For the last integral, let us use the substitution $x^2 = t$, so that $2x\, dx = dt$. This gives the following

$$I_4 = \frac{1}{2} \int t e^t dt.$$

Now, by using the integration by parts (assume $u = t, dv = e^t dt$), we get

$$I_4 = \frac{1}{2} \left( te^t - \int e^t dt \right) = \frac{1}{2} \left( te^t - e^t \right) = \frac{e^t}{2}(t - 1) + C = \frac{e^{x^2}}{2}(x^2 - 1) + C,$$

where $C \in \mathbb{R}$.

$$\boxed{I_5 = \int_1^2 \frac{dx}{x\sqrt{x^2 - 1}}}$$

**Solution.**

Let us use the substitution $1/x = t$, so that $x = 1/t$ and $dx = (-1/t^2)dt$. If $x = 1$, then $t = 1$, and if $x = 2$, then $t = 1/2$. With these new values, we get the following

$$I_5 = -\int_1^{1/2} \frac{t}{\sqrt{\frac{1}{t^2} - 1}} \frac{dt}{t^2} = \int_{1/2}^1 \frac{dt}{t\frac{\sqrt{1-t^2}}{t}} = \int_{1/2}^1 \frac{dt}{\sqrt{1 - t^2}} = [\arcsin t]_{1/2}^1$$

$$= \arcsin(1) - \arcsin(1/2) = \frac{\pi}{2} - \frac{\pi}{6} = \frac{\pi}{3}.$$

$$\boxed{I_6 = \int_1^\infty \frac{dx}{x(x^2 + 1)}}$$

**Solution.**

For the given integral we can write the following

$$I_6 = \lim_{a \to \infty} \int_1^a \frac{1 + x^2 - x^2}{x(1 + x^2)} = \lim_{a \to \infty} \int_1^a \left( \frac{1}{x} - \frac{x}{1 + x^2} \right) dx$$

$$= \lim_{a \to \infty} \left[ \ln x - \frac{1}{2} \ln(1 + x^2) \right]_1^a = \lim_{a \to \infty} \left[ \ln \left( \frac{x}{\sqrt{1 + x^2}} \right) \right]_1^a$$

$$= \lim_{a \to \infty} \left( \ln \left( \frac{a}{\sqrt{1 + a^2}} \right) - \ln \left( \frac{1}{\sqrt{2}} \right) \right)$$

$$= \lim_{a \to \infty} \left( \ln \left( \frac{a}{a\sqrt{1 + \frac{1}{a^2}}} \right) + \ln(\sqrt{2}) \right)$$

$$= \ln(\sqrt{2}) = \frac{\ln 2}{2}.$$

$$\boxed{I_7 = \int \cosh^{-1} x \, dx}$$

**Solution.**

For the given integral we can write the following

$$I_7 = \int \operatorname{arcosh} x \, dx = \int \ln \left( x + \sqrt{x^2 - 1} \right) dx.$$

Now, for the last integral, by using the integration by parts, let us assume

$$u = \ln \left( x + \sqrt{x^2 - 1} \right) \implies du = \frac{dx}{\sqrt{x^2 - 1}}, \quad dv = dx \implies v = x.$$

Thus, we have

$$I_7 = x \ln \left( x + \sqrt{x^2 - 1} \right) - \int \frac{x}{\sqrt{x^2 - 1}} dx$$

$$= x \ln \left( x + \sqrt{x^2 - 1} \right) - \sqrt{x^2 - 1} + C$$

$$= x \operatorname{arcosh} x - \sqrt{x^2 - 1} + C,$$

where $C \in \mathbb{R}$.

$$I_8 = \int_{-\infty}^{\infty} e^{-2x^2 - 5x - 3} dx$$

**Solution.**

We have

$$-2x^2 - 5x - 3 = -2 \left( x^2 + \frac{5}{2}x \right) - 3 = -2 \left( x^2 + \frac{5}{2}x + \frac{25}{16} - \frac{25}{16} \right) - 3$$

$$= -2 \left( x + \frac{5}{4} \right)^2 + \frac{1}{8}.$$

Thus, for the given integral, we get

$$I_8 = \int_{-\infty}^{\infty} e^{-2\left( x + \frac{5}{4} \right)^2 + \frac{1}{8}} dx = e^{\frac{1}{8}} \int_{-\infty}^{\infty} e^{-2\left( x + \frac{5}{4} \right)^2} dx.$$

Let us use the substitution $x + \frac{5}{4} = t$, so that $dx = dt$. If $x \to -\infty$, then $t \to -\infty$, and if $x \to \infty$, then $t \to \infty$. This gives the following

$$I_8 = \frac{e^{\frac{1}{8}}}{\sqrt{2}} \int_{-\infty}^{\infty} e^{-t^2} dt = e^{\frac{1}{8}} \sqrt{\frac{\pi}{2}}.$$

Where we used the Gaussian integral (1.14).

$$I_9 = \int \sin \left( \sqrt{x} \right) dx$$

**Solution.**

Let us use the substitution $\sqrt{x} = t$, so that $x = t^2$, and $dx = 2t dt$. This gives the following

$$I_9 = 2 \int t \sin t \, dt.$$

Now, by using the integration by parts (assume $u = t, dv = \sin t dt$), we get

$$I_9 = 2 \left( -t \cos t + \sin t \right) + C = -2\sqrt{x} \cos \left( \sqrt{x} \right) + 2 \sin \left( \sqrt{x} \right) + C,$$

where $C \in \mathbb{R}$.

$$I_{10} = \int_0^\infty \frac{dx}{\left(x + \frac{1}{x}\right)^2}$$

## Solution.

Let us calculate the indefinite integral

$$I := \int \frac{dx}{\left(x + \frac{1}{x}\right)^2} = \int \frac{dx}{x^2 \left(1 + \frac{1}{x^2}\right)^2}.$$

For the last integral, let us use the substitution $1/x = t$, so that $x = 1/t$, and $dx = \left(-1/t^2\right) dt$. This gives the following

$$I = -\int \frac{dt}{t^2 \frac{1}{t^2} (1 + t^2)^2} = -\int \frac{dt}{(1 + t^2)^2} = -\int \frac{1 + t^2 - t^2}{(1 + t^2)^2} dt$$

$$= -\int \left(\frac{1}{1 + t^2} - \frac{t^2}{(1 + t^2)^2}\right) dt = \underbrace{\int t \frac{t}{(1 + t^2)^2} dt}_{:=J} - \int \frac{dt}{1 + t^2}.$$

For the integral $J$, by using the integration by parts, let us assume

$$u = t \Longrightarrow du = dt, \ dv = \frac{t}{(1 + t^2)^2} dt \Longrightarrow v = \int t(1 + t^2)^{-2} dt = -\frac{1}{2(1 + t^2)}.$$

Thus, we have

$$I = -\frac{t}{2(1 + t^2)} + \frac{1}{2} \int \frac{dt}{1 + t^2} - \int \frac{dt}{1 + t^2} = -\frac{t}{2(1 + t^2)} - \frac{1}{2} \int \frac{dt}{1 + t^2}$$

$$= -\frac{t}{2(1 + t^2)} - \frac{1}{2} \arctan t + C = -\frac{1}{2x \left(\frac{1 + x^2}{x^2}\right)} - \frac{1}{2} \arctan\left(\frac{1}{x}\right) + C$$

$$= -\frac{x}{2(1 + x^2)} - \frac{1}{2} \arctan\left(\frac{1}{x}\right) + C,$$

where $C \in \mathbb{R}$.

Therefore, for the given integral $I_{10}$, we get

$$I_{10} = \lim_{\substack{a \to 0^+ \\ b \to \infty}} \left[-\frac{x}{2(1 + x^2)} - \frac{1}{2} \arctan\left(\frac{1}{x}\right)\right]_a^b$$

$$= \lim_{\substack{a \to 0^+ \\ b \to \infty}} \left(-\frac{b}{2(1 + b^2)} - \frac{1}{2} \arctan\left(\frac{1}{b}\right)\right)$$

$$- \lim_{\substack{a \to 0^+ \\ b \to \infty}} \left(-\frac{a}{2(1 + a^2)} - \frac{1}{2} \arctan\left(\frac{1}{a}\right)\right) = \frac{\pi}{4}.$$

$$I_{11} = \int \frac{(2+x)e^{-x}}{x^3}\,dx$$

**Solution.**

For the given integral we can write the following

$$I_{11} = 2\underbrace{\int \frac{e^{-x}}{x^3}\,dx}_{:=J} + \int \frac{e^{-x}}{x^2}\,dx.$$

For the integral $J$, by using the integration by parts, let us assume

$$u = e^{-x} \implies du = -e^{-x}dx, \quad dv = \frac{dx}{x^3} \implies v = -\frac{1}{2x^2}.$$

Thus, we have

$$I_{11} = 2\left(-\frac{e^{-x}}{2x^2} - \int \frac{e^{-x}}{2x^2}\,dx\right) + \int \frac{e^{-x}}{x^2}\,dx = -\frac{e^{-x}}{x^2} + C,$$

where $C \in \mathbb{R}$.

$$I_{12} = \int_0^1 \frac{dx}{\sqrt{x(1-x)}}$$

**Solution.**

For the given integral we can write the following

$$I_{12} = \lim_{\varepsilon \to 0^+} \int_\varepsilon^{1-\varepsilon} \frac{dx}{\sqrt{-x^2+x}} = \lim_{\varepsilon \to 0^+} \int_\varepsilon^{1-\varepsilon} \frac{dx}{\sqrt{-\left(x^2 - x + \frac{1}{4} - \frac{1}{4}\right)}}$$

$$= \lim_{\varepsilon \to 0^+} \int_\varepsilon^{1-\varepsilon} \frac{dx}{\sqrt{\frac{1}{4} - \left(x - \frac{1}{2}\right)^2}} = \lim_{\varepsilon \to 0^+} \left[\arcsin\left(\frac{x - \frac{1}{2}}{\frac{1}{2}}\right)\right]_\varepsilon^{1-\varepsilon}$$

$$= \lim_{\varepsilon \to 0^+} \left[\arcsin(2x - 1)\right]_\varepsilon^{1-\varepsilon} = \lim_{\varepsilon \to 0^+} \left(\arcsin(1 - 2\varepsilon) - \arcsin(2\varepsilon - 1)\right)$$

$$= \arcsin(1) - \arcsin(-1) = 2\arcsin(1) = \pi.$$

$$I_{13} = \int_0^\infty \frac{\tanh x}{e^x}\,dx$$

**Solution.**

For the given integral we can write the following

$$I_{13} = \int_0^\infty \frac{e^x - e^{-x}}{e^x (e^x + e^{-x})} dx = \int_0^\infty \frac{1 - e^{-2x}}{e^x + e^{-x}} dx = \int_0^\infty \frac{1 - e^{-2x}}{e^x (1 + e^{-2x})} dx$$

$$= \int_0^\infty \frac{1 - e^{-2x}}{1 + e^{-2x}} e^{-x} dx = \lim_{a \to \infty} \int_0^a \frac{1 - e^{-2x}}{1 + e^{-2x}} e^{-x} dx.$$

For the last integral, let us use the substitution $e^{-x} = t$, so that $e^{-x} dx = -dt$. If $x = 0$, then $t = 1$, and if $x = a$, then $t = e^{-a}$. This gives the following

$$I_{13} = -\lim_{a \to \infty} \int_1^{e^{-a}} \frac{1 - t^2}{1 + t^2} dt$$

$$= -\lim_{a \to \infty} \int_1^{e^{-a}} \left(1 - \frac{2}{1 + t^2}\right) dt$$

$$= \lim_{a \to \infty} [t - 2\arctan t]_1^{e^{-a}}$$

$$= \lim_{a \to \infty} \left(e^{-a} - 2\arctan\left(e^{-a}\right)\right) - (1 - 2\arctan(1))$$

$$= \frac{\pi}{2} - 1.$$

$$\boxed{I_{14} = \int_0^{\pi/2} \sqrt{1 + \sin x} dx}$$

**Solution.**

For the given integral we have

$$I_{14} = \int_0^{\pi/2} \sqrt{1 + 2\sin(x/2)\cos(x/2)} dx$$

$$= \int_0^{\pi/2} \sqrt{\sin^2(x/2) + \cos^2(x/2) + 2\sin(x/2)\cos(x/2)} dx$$

$$= \int_0^{\pi/2} \sqrt{(\sin(x/2) + \cos(x/2))^2} dx = \int_0^{\pi/2} (\sin(x/2) + \cos(x/2)) dx$$

$$= [-2\cos(x/2) + 2\sin(x/2)]_0^{\pi/2} = \left(-\sqrt{2} + \sqrt{2}\right) - (-2) = 2.$$

$$\boxed{\lim_{n \to \infty} I_n =? \quad \text{where} \quad I_1 = \int_0^1 \frac{dx}{1 + \sqrt{x}}, \quad I_2 = \int_0^1 \frac{dx}{1 + \frac{1}{1 + \sqrt{x}}}, \quad \cdots}$$

**Solution.**

Let us denote $f_1(x) := \frac{1}{1+\sqrt{x}}$, for the integrand in $I_1$. Thus, for $I_2$ we have

$$I_2 = \int_0^1 \frac{dx}{1 + \frac{1}{1+\sqrt{x}}} = \int_0^1 \frac{dx}{1 + f_1(x)} = \int_0^1 \frac{1+\sqrt{x}}{2+\sqrt{x}} dx,$$

and the integrand in $I_2$ is $f_2(x) := \frac{1+\sqrt{x}}{2+\sqrt{x}}$.

For $I_3$, we have

$$I_3 = \int_0^1 \frac{dx}{1 + \frac{1}{1+\frac{1}{1+\sqrt{x}}}} = \int_0^1 \frac{dx}{1 + f_2(x)} = \int_0^1 \frac{dx}{1 + \frac{1+\sqrt{x}}{2+\sqrt{x}}} = \int_0^1 \frac{2+\sqrt{x}}{3+2\sqrt{x}} dx,$$

and the integrand in $I_3$ is $f_3(x) := \frac{2+\sqrt{x}}{3+2\sqrt{x}}$.

For $I_4$, we have

$$I_4 = \int_0^1 \frac{dx}{1 + f_3(x)} = \int_0^1 \frac{dx}{1 + \frac{2+\sqrt{x}}{3+2\sqrt{x}}} = \int_0^1 \frac{3+2\sqrt{x}}{5+3\sqrt{x}} dx,$$

and the integrand in $I_4$ is $f_4(x) := \frac{3+2\sqrt{x}}{5+3\sqrt{x}}$.

Thus, we have the following integrands in $I_1, I_2, I_3, \dots$

$$f_1(x) := \frac{1}{1+\sqrt{x}}, \quad f_2(x) := \frac{1+\sqrt{x}}{2+\sqrt{x}}, \quad f_3(x) := \frac{2+\sqrt{x}}{3+2\sqrt{x}},$$

$$f_4(x) := \frac{3+2\sqrt{x}}{5+3\sqrt{x}}, \quad f_5(x) := \frac{5+3\sqrt{x}}{8+5\sqrt{x}}, \quad f_6(x) = \frac{8+5\sqrt{x}}{13+8\sqrt{x}}, \quad \dots$$

Let us set

$$a_0 := 1, \quad a_1 := 1+\sqrt{x}, \quad a_2 := 2+\sqrt{x}, \quad a_3 := 3+2\sqrt{x},$$
$$a_4 := 5+3\sqrt{x}, \quad a_5 := 8+5\sqrt{x}, \quad a_6 := 13+8\sqrt{x}, \quad \dots \tag{10.1}$$

Then we get

$$f_1(x) = \frac{a_0}{a_1}, \quad f_2(x) = \frac{a_1}{a_2}, \quad f_3(x) = \frac{a_2}{a_3}, \quad f_4(x) = \frac{a_3}{a_4},$$
$$f_5(x) = \frac{a_4}{a_5}, \dots, \quad f_n(x) = \frac{a_{n-1}}{a_n}, \quad \dots$$

We note that,

$$a_2 = a_0 + a_1, \quad a_3 = a_1 + a_2, \quad a_4 = a_2 + a_3, \quad a_5 = a_3 + a_4, \quad \dots$$

By induction, we have the following recurrence relation

$$a_n = a_{n-2} + a_{n-1}, \quad \forall n \geqslant 2.$$

Thus the sequence $\{a_n\}_{n \geq 0}$ (see (10.1)) is a Fibonacci sequence. Therefore, we have $\lim_{n \to \infty} \frac{a_n}{a_{n-1}} = \varphi = \frac{1+\sqrt{5}}{2}$, which is the golden ratio.

As a result, we find

$$I_n = \int_0^1 f_n(x)dx = \int_0^1 \frac{a_{n-1}}{a_n}dx.$$

Thus

$$\lim_{n \to \infty} I_n = \lim_{n \to \infty} \int_0^1 \frac{a_{n-1}}{a_n}dx = \int_0^1 \left(\lim_{n \to \infty} \frac{a_{n-1}}{a_n}\right) dx = \int_0^1 \frac{1}{\varphi}dx$$

$$= \int_0^1 \frac{2}{1+\sqrt{5}}dx = \frac{2}{1+\sqrt{5}} = \frac{2(1-\sqrt{5})}{1-5} = \frac{\sqrt{5}-1}{2}.$$

$$\boxed{I_{16} = \int_{-\infty}^{\infty} \frac{\sin^2(x + \pi/4)}{e^{x^2}}dx}$$

**Solution.**

We know that

$$\sin\left(x + \frac{\pi}{4}\right) = \sin x \sin\left(\frac{\pi}{4}\right) + \cos x \cos\left(\frac{\pi}{4}\right).$$

Thus,

$$\sin^2\left(x + \frac{\pi}{4}\right) = \frac{1}{2}\left(\sin^2 x + \cos^2 x + 2\sin x \cos x\right) = \frac{1}{2}(1 + \sin(2x)).$$

Therefore, for the given integral we get

$$I_{16} = \frac{1}{2}\int_{-\infty}^{\infty}(1 + \sin(2x))e^{-x^2}dx$$

$$= \frac{1}{2}\int_{-\infty}^{\infty} e^{-x^2}dx + \frac{1}{2}\int_{-\infty}^{\infty}\sin(2x)e^{-x^2}dx.$$

The function $f(x) := \sin(2x)e^{-x^2}$ is an odd function in $\mathbb{R}$ (i.e. $f(-x) = -f(x), \forall x \in \mathbb{R}$). Thus,

$$\int_{-\infty}^{\infty}\sin(2x)e^{-x^2}dx = 0.$$

Therefore, for the given integral $I_{16}$, we get

$$I_{16} = \frac{1}{2}\int_{-\infty}^{\infty} e^{-x^2}dx = \frac{\sqrt{\pi}}{2}.$$

Where we used the Gaussian integral (1.14).

$$I_{17} = \int_{-\infty}^{\infty} 3x^2 \left(x^3 + 1\right)^2 e^{-x^6 - 2x^3} dx$$

**Solution.**

For the given integral we can write the following

$$I_{17} = \lim_{a \to \infty} \int_{-a}^{a} 3x^2 \left(x^3 + 1\right)^2 e^{-x^6 - 2x^3} dx.$$

Let us use the substitution $x^3 + 1 = t$, so that $3x^2 dx = dt$.

$$(x^3 + 1)^2 = t^2 \implies x^6 + 2x^3 + 1 = t^2 \implies -x^6 - 2x^3 = 1 - t^2.$$

If $x = -a$, then $t = -a$, and if $x = a$, then $t = a$. With these new values, we get the following

$$I_{17} = \lim_{a \to \infty} \int_{-a}^{a} t^2 e^{1-t^2} dt = e \lim_{a \to \infty} \int_{-a}^{a} t^2 e^{-t^2} dt = \lim_{a \to \infty} \int_{-a}^{a} te^{-t^2} t \, dt.$$

Now, by using the integration by parts, let us assume

$$u = t \implies du = dt, \quad dv = te^{-t^2} dt \implies v = -\frac{1}{2}e^{-t^2}.$$

Thus, we have

$$I_{17} = e \lim_{a \to \infty} \left( \left[ -\frac{1}{2}te^{-t^2} \right]_{-a}^{a} + \frac{1}{2} \int_{-a}^{a} e^{-t^2} dt \right)$$

$$= e \lim_{a \to \infty} \left( -\frac{a}{2}e^{-a^2} - \frac{a}{2}e^{-a^2} + \frac{1}{2} \int_{-a}^{a} e^{-t^2} dt \right)$$

$$= e \lim_{a \to \infty} \left( -ae^{-a^2} \right) + \frac{e}{2} \int_{-\infty}^{\infty} e^{-t^2} dt = \frac{e\sqrt{\pi}}{2}.$$

Where we used the Gaussian integral (1.14).

$$I_{18} = \int_{0}^{\pi/2} \frac{dx}{1 + \tan^{2017} x}$$

**Solution.**

For the given integral we can write the following

$$I_{18} = \int_{0}^{\pi/2} \frac{dx}{1 + \tan^{2017} x} = \int_{0}^{\pi/2} \frac{\cos^{2017} x}{\sin^{2017} x + \cos^{2017} x} dx. \qquad (10.2)$$

For the last integral, let us use the substitution $\frac{\pi}{2} - t = x$, so that $dx = -dt$. If $x = 0$, then $t = \pi/2$, and if $x = \pi/2$, then $t = 0$. With these new values, we get the following

$$I_{18} = -\int_{\pi/2}^{0} \frac{\sin^{2017} t}{\cos^{2017} t + \sin^{2017} t} dt = \int_{0}^{\pi/2} \frac{\sin^{2017} t}{\cos^{2017} t + \sin^{2017} t} dt. \qquad (10.3)$$

By summing (10.2) and (10.3), we find

$$2I_{18} = \int_{0}^{\pi/2} \frac{\sin^{2017} t + \cos^{2017} t}{\cos^{2017} t + \sin^{2017} t} dt = \int_{0}^{\pi/2} dt = \frac{\pi}{2}.$$

Thus,

$$I_{18} = \frac{\pi}{4}.$$

$$\boxed{I_{19} = \int e^{2x} \cos(3x) dx}$$

**Solution.**

The given integral has the following form

$$\int e^{ax} \cos(bx) dx = \frac{e^{ax}}{a^2 + b^2} (a \cos(bx) + b \sin(bx)) + C,$$

where $a = 2, b = 3$ (see (1.1)). Therefore, we get

$$I_{19} = \frac{e^{2x}}{13} (3 \sin(3x) + 2 \cos(3x)) + C,$$

where $C \in \mathbb{R}$.

$$\boxed{I_{20} = \int ((\cos x)^{\cos x + 1} \tan x)(1 + \ln(\cos x)) dx}$$

**Solution.**

For the given integral we can write the following

$$I_{20} = \int \cos x (\cos x)^{\cos x} \frac{\sin x}{\cos x} (1 + \ln(\cos x)) \, dx$$

$$= \int (\cos x)^{\cos x} (1 + \ln(\cos x)) \sin x dx.$$

For the last integral, let us use the substitution $\cos x = t$, so that $\sin x \, dx = -dt$. This gives the following

$$I_{20} = -\int t^t \left(1 + \ln t\right) dt = -\int e^{t \ln t} \left(1 + \ln t\right) dt.$$

Now, let us use the substitution $t \ln t = y$, so that $(1 + \ln t) \, dt = dy$. This gives the following

$$I_{20} = -\int e^y dy = -e^y + C = -e^{t \ln t} + C = -t^t + C = -(\cos x)^{\cos x} + C.$$

where $C \in \mathbb{R}$.

# Chapter 11

# The Solutions to the 2018 MIT Integration Bee, Qualifying Test

$$I_1 = \int \frac{e^x}{e^x + 2} dx$$

**Solution.**

This integral is very simple, where $(e^x + 2)' = e^x$. Thus, $I_1 = \ln(e^x + 2) + C$, where $C \in \mathbb{R}$.

$$I_2 = \int \sqrt{x \sqrt[3]{x \sqrt[4]{x \sqrt[5]{x \ldots}}}} \, dx$$

**Solution.**

For the given integral we can write the following

$$I_2 = \int x^{\frac{1}{2}} \left( \sqrt[3]{x \sqrt[4]{x \sqrt[5]{x \sqrt[6]{x \ldots}}}} \right)^{1/2} dx$$

$$= \int x^{\frac{1}{2}} x^{\frac{1}{3}\frac{1}{2}} \left( \sqrt[4]{x \sqrt[5]{x \sqrt[6]{x \ldots}}} \right)^{\frac{1}{3}\frac{1}{2}} dx$$

$$= \int x^{\frac{1}{2}} x^{\frac{1}{3}\frac{1}{2}} x^{\frac{1}{4}\frac{1}{3}\frac{1}{2}} \left( \sqrt[5]{x \sqrt[6]{x \ldots}} \right)^{\frac{1}{4}\frac{1}{3}\frac{1}{2}} dx$$

$$= \int x^{\frac{1}{2!}} x^{\frac{1}{3!}} x^{\frac{1}{4!}} x^{\frac{1}{5!}} \cdots dx = \int x^{\sum_{n=2}^{\infty} \frac{1}{n!}} dx$$

$$= \int x^{\left(\sum_{n=0}^{\infty} \frac{1}{n!}\right) - 2} dx = \int x^{e-2} dx = \frac{x^{e-1}}{e-1} + C,$$

where $C \in \mathbb{R}$.

$$\boxed{I_3 = \int_0^{2018\pi} |\sin(2018x)| \, dx}$$

**Solution.**

Let us use the substitution $2018x = t$, so that $2018dx = dt$. If $x = 0$, then $t = 0$, and if $x = 2018\pi$, then $t = 2018^2\pi$. With these new values, we get the following

$$I_3 = \frac{1}{2018} \int_0^{2018^2\pi} |\sin t| \, dt.$$

The function $f(t) := |\sin t|$ is periodic, with period $P = \pi$ (see Fig. 11.1). Thus, we can write the following

$$I_3 = \frac{2018^2}{2018} \int_0^{\pi} |\sin t| \, dt = 2018 \int_0^{\pi} \sin t \, dt = -2018 \left[\cos t\right]_0^{\pi}$$

$$= -2018(-2) = 4036.$$

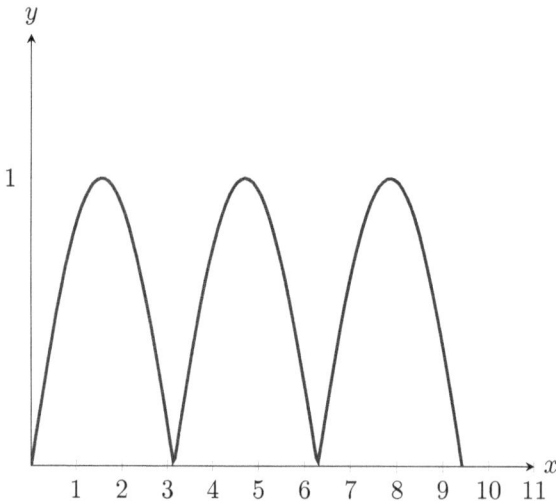

Figure 11.1:   The graph of $f(x) = |\sin x|$ in $[0, 3\pi]$. There are 3 hills above the $x$-axis. In $[0, 2018^2\pi]$ it will be $2018^2$ hills above the $x$-axis.

**Remark.** In general, let $a \in \mathbb{N}$, then we can find

$$\int_0^{a\pi} |\sin(a\pi)| \, dx = \frac{1}{a} \int_0^{a^2\pi} |\sin x| \, dx = \frac{a^2}{a} \int_0^{\pi} |\sin x| dx$$

$$= a \int_0^{\pi} \sin x dx = 2a.$$

Similarly, we also find

$$\int_0^{a\pi} |\cos(ax)| \, dx = 2a.$$

$$\boxed{I_4 = \int \frac{dx}{\tan x + \cot x}}$$

**Solution.**

For the given integral we can write the following

$$I_4 = \int \frac{dx}{\frac{\sin x}{\cos x} + \frac{\cos x}{\sin x}} = \int \frac{dx}{\frac{\sin^2 x + \cos^2 x}{\sin x \cos x}} = \int \sin x \cos x dx = \frac{1}{2} \int \sin(2x) dx$$

$$= -\frac{1}{4} \cos(2x) + C_1 \quad \text{(where } C_1 \in \mathbb{R}\text{)}$$

$$= -\frac{1}{4} \left(1 - 2\sin^2 x\right) + C_1 = -\frac{1}{4} + \frac{1}{2}\sin^2 x + C_1 = \frac{1}{2}\sin^2 x + C,$$

where $C = C_1 - \frac{1}{4} \in \mathbb{R}$.

$$\boxed{I_5 = \int \frac{x^5}{2 + x^{12}} dx}$$

**Solution.**

For the given integral we can write the following

$$I_5 = \int \frac{x^5}{2 + (x^6)^2} dx.$$

For the last integral, let us use the substitution $x^6 = t$, so that $6x^5 dx = dt$. This gives the following

$$I_5 = \frac{1}{6} \int \frac{dx}{2 + t^2} = \frac{1}{6\sqrt{2}} \arctan\left(\frac{t}{\sqrt{2}}\right) + C = \frac{1}{6\sqrt{2}} \arctan\left(\frac{x^6}{\sqrt{2}}\right) + C,$$

where $C \in \mathbb{R}$.

$$\boxed{I_6 = \int (\cos x \cosh x + \sin x \sinh x) \, dx}$$

**Solution 1.**

This integral is very simple, where

$$I_6 = \int (\sin x \cosh x)' \, dx = \sin x \cosh x + C,$$

where $C \in \mathbb{R}$.

**Solution 2.**

We have

$$I_6 = \underbrace{\int \cos x \cosh x dx}_{:=J} + \int \sin x \sinh x dx.$$

For the integral $J$, by using the integration by parts, let us assume

$$u = \cosh x \Longrightarrow du = \sinh x dx, \quad dv = \cos x dx \Longrightarrow v = \sin x.$$

Thus, we have

$$I_6 = \cosh x \sin x - \int \sin x \sinh x dx + \int \sin x \sinh x dx = \cosh x \sin x + C,$$

where $C \in \mathbb{R}$.

$$\boxed{I_7 = \int \frac{e^x + \cos x}{e^x + \sin x} dx}$$

**Solution.**

This integral is very simple, where

$$I_7 = \int \frac{(e^x + \sin x)'}{e^x + \sin x} dx = \ln |e^x + \sin x| + C,$$

and $C \in \mathbb{R}$.

$$\boxed{I_8 = \int \sin(\cos(\sin x)) \sin(\sin x) \cos x \, dx}$$

**Solution.**

Let us use the substitution $\sin x = t$, so that $\cos x dx = dt$. This gives the following

$$I_8 = \int \sin(\cos t) \sin t \, dt.$$

Now, let us use the substitution $\cos t = u$, so that $\sin t\, dt = -du$. This gives the following

$$I_8 = -\int \sin u\, du = \cos u + C = \cos(\cos t) + C = \cos\left(\cos(\sin x)\right) + C,$$

where $C \in \mathbb{R}$.

$$\boxed{I_9 = \int \frac{dx}{1 + \sin x}}$$

**Solution 1.**

Let us use the substitution $\tan\left(\frac{x}{2}\right) = t$, so that $\sin x = \frac{2t}{1+t^2}$, and $dx = \frac{2}{1+t^2}dt$. This gives the following

$$I_8 = \int \frac{\frac{2dt}{1+t^2}}{1 + \frac{2t}{1+t^2}} = 2\int \frac{dt}{t^2 + 2t + 1} = 2\int \frac{dt}{(t+1)^2} = -\frac{2}{1+t} + C$$

$$= -\frac{2}{1 + \tan(x/2)} + C,$$

where $C \in \mathbb{R}$.

**Solution 2.**

For the given integral we can write the following

$$I_9 = \int \frac{dx}{\underbrace{\sin^2\left(\frac{x}{2}\right) + \cos^2\left(\frac{x}{2}\right)}_{=1} + \underbrace{2\sin\left(\frac{x}{2}\right)\cos\left(\frac{x}{2}\right)}_{=\sin x}}$$

$$= \frac{1}{2}\int \frac{dx}{\left(\frac{1}{\sqrt{2}}\sin\left(\frac{x}{2}\right) + \frac{1}{\sqrt{2}}\cos\left(\frac{x}{2}\right)\right)^2}$$

$$= \frac{1}{2}\int \frac{dx}{\left(\sin\left(\frac{\pi}{4}\right)\sin\left(\frac{x}{2}\right) + \cos\left(\frac{\pi}{4}\right)\cos\left(\frac{x}{2}\right)\right)^2}$$

$$= \frac{1}{2}\int \frac{dx}{\cos^2\left(\frac{x}{2} - \frac{\pi}{4}\right)}.$$

For the last integral, let us use the substitution $\frac{x}{2} - \frac{\pi}{4} = t$, so that $dx = 2dt$. This gives the following

$$I_9 = \int \frac{dt}{\cos^2 t} = \tan t + C = \tan\left(\frac{x}{2} - \frac{\pi}{4}\right) + C,$$

where $C \in \mathbb{R}$.

$$I_{10} = \int \frac{\cos x}{1 - \cos(2x)} \, dx$$

**Solution.**

For the given integral we can write the following

$$I_{10} = \int \frac{\cos x}{1 - (\cos^2 x - \sin^2 x)} \, dx = \int \frac{\cos x}{1 - (1 - \sin^2 x) + \sin^2 x} \, dx$$

$$= \int \frac{\cos x}{2 \sin^2 x} \, dx = \frac{1}{2} \int \cos x (\sin x)^{-2} \, dx = -\frac{1}{2 \sin x} + C,$$

where $C \in \mathbb{R}$.

$$I_{11} = \int e^x \left( \frac{1}{x} + \ln x \right) dx$$

**Solution 1.**

We have

$$e^x \left( \frac{1}{x} + \ln x \right) = \frac{e^x}{x} + e^x \ln x,$$

and

$$(e^x \ln x)' = \frac{e^x}{x} + e^x \ln x.$$

Thus, we have

$$I_{11} = \int \left( \frac{e^x}{x} + e^x \ln x \right) dx = \int (e^x \ln x)' \, dx = e^x \ln x + C,$$

where $C \in \mathbb{R}$.

**Solution 2.**

By using the integration by parts, let us assume

$$u = \frac{1}{x} + \ln x \implies du = \left( -\frac{1}{x^2} + \frac{1}{x} \right) dx, \quad dv = e^x dx \implies v = e^x.$$

Thus, we have

$$I_{11} = \left( \frac{1}{x} + \ln x \right) e^x - \int e^x \left( -\frac{1}{x^2} + \frac{1}{x} \right) dx$$

$$= \left( \frac{1}{x} + \ln x \right) e^x + \int \frac{e^x}{x^2} dx - \underbrace{\int \frac{e^x}{x} dx}_{:=J}.$$

Now, for the integral $J$, by using the integration by parts, let us assume

$$u = \frac{1}{x} \implies du = du = -\frac{dx}{x^2}, \quad dv = e^x dx \implies v = e^x.$$

Thus, we have

$$J = \frac{e^x}{x} + \int \frac{e^x}{x^2} dx.$$

Therefore, for the given integral $I_{11}$, we get

$$I_{11} = \left( \frac{1}{x} + \ln x \right) e^x + \int \frac{e^x}{x^2} dx - \frac{e^x}{x} - \int \frac{e^x}{x^2} dx = e^x \ln x + C,$$

where $C \in \mathbb{R}$.

$$\boxed{I_{12} = \int \tanh^2 x\, dx}$$

**Solution.**

For the given integral we can write the following

$$I_{12} = \int \frac{\sinh^2 x}{\cosh^2 x} dx = \int \frac{\cosh^2 x - 1}{\cosh^2 x} dx = \int \left( 1 - \frac{1}{\cosh^2 x} \right) dx$$
$$= x - \tanh x + C,$$

where $C \in \mathbb{R}$.

$$\boxed{I_{13} = \int \frac{2017 x^{2016} + 2018 x^{2017}}{1 + x^{4034} + 2 x^{4035} + x^{4036}} dx}$$

**Solution.**

For the given integral we can write the following

$$I_{13} = \int \frac{2017 x^{2016} + 2018 x^{2017}}{1 + (x^{2017} + x^{2018})^2} dx.$$

Let us use the substitution $x^{2017} + x^{2018} = t$, so that $\left(2017 x^{2016} + 2018 x^{2017}\right) dx = dt$. This gives the following

$$I_{13} = \int \frac{dt}{1 + t^2} = \arctan t + C = \arctan \left( x^{2017} + x^{2018} \right) + C,$$

where $C \in \mathbb{R}$.

**Remark.** We can generalize the integral $I_{13}$, to the following form[1]

$$J_a = \int \frac{ax^{a-1} + (a+1)x^a}{1 + x^{2a} + 2x^{2a+1} + x^{2a+2}} dx = \int \frac{ax^{a-1} + (a+1)x^a}{1 + (x^a + x^{a+1})^2} dx, \quad (a \in \mathbb{R}).$$

By using the substitution $x^a + x^{a+1} = t$, we get

$$J_a = \int \frac{dt}{1 + t^2} = \arctan t + C = \arctan(x^a + x^{a+1}) + C,$$

where $C \in \mathbb{R}$.

When we take $a = 2017$, we get the integral $I_{13} = J_{2017}$.

$$I_{14} = \int \frac{\sin(2x) - \sin^2 x}{\cos(2x) - \cos^2 x} dx$$

**Solution.**

For the given integral we can write the following

$$I_{14} = \int \frac{\sin(2x) - \left(\frac{1 - \cos(2x)}{2}\right)}{\cos(2x) - \left(\frac{1 + \cos(2x)}{2}\right)} dx$$

$$= \int \frac{2\sin(2x) - 1 + \cos(2x)}{2\cos(2x) - 1 - \cos(2x)} dx$$

$$= \int \frac{2\sin(2x) - 1 + \cos(2x)}{-1 + \cos(2x)} dx$$

$$= \int \frac{2\sin(2x)}{-1 + \cos(2x)} dx + \int \frac{-1 + \cos(2x)}{-1 + \cos(2x)} dx$$

$$= -\int \frac{-2\sin(2x)}{-1 + \cos(2x)} dx + \int dx$$

$$= x - \ln|-1 + \cos(2x)| + C_1 \quad \text{(where } C_1 \in \mathbb{R}\text{)}$$

$$= x - \ln|1 - \cos(2x)| + C_1$$

$$= x - \ln|2\sin^2 x| + C_1 = x - \ln 2 - 2\ln(\sin x) + C_1$$

$$= x - 2\ln(\sin x) + C,$$

where $C = C_1 - \ln 2 \in \mathbb{R}$.

$$I_{15} = \int \frac{dx}{x^{25/25}x^{16/25} + x^{9/25}}$$

**Solution 1.**

---

[1] The idea of this generalization taken from https://phildonnia.quora.com/Solutions-to-the-2018-MIT-Integration-Bee-Qualifying-Exam

For the given integral we can write the following

$$I_{15} = \int \frac{dx}{xx^{16/25} + x^{9/25}} = \int \frac{dx}{x^{9/25}\left(1 + x^{16/25}x^{16/25}\right)}$$

$$= \int \frac{x^{-9/25}}{1 + \left(x^{16/25}\right)^2}dx.$$

For the last integral, let us use the substitution $x^{16/25} = t$, so that $x^{-9/25}dx = \frac{25}{16}dt$. This gives the following

$$I_{15} = \frac{25}{16}\int \frac{dt}{1 + t^2} = \frac{25}{16}\arctan t + C = \frac{25}{16}\arctan\left(x^{16/25}\right) + C,$$

where $C \in \mathbb{R}$.

**Solution 2.**

For the given integral we can write the following

$$I_{15} = \int \frac{dx}{x^{41/25} + x^{9/25}}.$$

Let us use the substitution $x = t^{25}$, so that $dx = 25t^{24}dt$. This gives the following

$$I_{15} = 25\int \frac{t^{24}}{t^{41} + t^9}dt = 25\int \frac{t^{24}}{t^9(1 + t^{32})}dt = 25\int \frac{t^{15}}{1 + t^{32}}dt$$

$$= 25\int \frac{t^{15}}{1 + (t^{16})^2}dt.$$

Let us use the substitution $t^{16} = u$, so that $16t^{15}dt = du$. This gives the following

$$I_{15} = \frac{25}{16}\int \frac{du}{1 + u^2} = \frac{25}{16}\arctan u + C = \frac{25}{16}\arctan(t^{16}) + C$$

$$= \frac{25}{16}\arctan\left(x^{16/25}\right) + C,$$

where $C \in \mathbb{R}$.

$$\boxed{I_{16} = \int_0^{\pi/2} \frac{\cos x}{2 - \cos^2 x}dx}$$

**Solution.**

For the given integral we can write the following

$$I_{16} = \int_0^{\pi/2} \frac{\cos x}{2 - (1 - \sin^2 x)}dx = \int_0^{\pi/2} \frac{\cos x}{1 + \sin^2 x}dx.$$

For the last integral, let us use the substitution $\sin x = t$, so that $\cos x \, dx = dt$. If $x = 0$, then $t = 0$, and if $x = \pi/2$, then $t = 1$. With these new values, we get the following

$$I_{16} = \int_0^1 \frac{dt}{1 + t^2} = [\arctan t]_0^1 = \arctan(1) - \arctan(0) = \frac{\pi}{4}.$$

$$I_{17} = \int \frac{dx}{(1 + x^2)^{3/2}}$$

**Solution 1.**

For the given integral we can write the following

$$I_{17} = \int \frac{dx}{(x^2)^{3/2} \left(1 + x^{-2}\right)^{3/2}} = \int \frac{x^{-3}}{\left(1 + x^{-2}\right)^{3/2}} dx.$$

For the last integral, let us use the substitution $x^{-2} = t$, so that $x^{-3} dx = -dt/2$. This gives the following

$$I_{17} = -\frac{1}{2} \int \frac{dt}{(1 + t)^{3/2}} = -\frac{1}{2} \int (1 + t)^{-3/2} dt = (1 + t)^{-1/2} + C$$

$$= \frac{1}{\sqrt{1 + t}} + C = \frac{1}{\sqrt{1 + \frac{1}{x^2}}} + C = \frac{x}{\sqrt{1 + x^2}} + C,$$

where $C \in \mathbb{R}$.

Note that, in our solution, we considered $x > 0$ (Be attentive, when the integral is definite!).

**Solution 2.**

For the given integral we can write the following

$$I_{17} = \int \frac{dx}{(1 + x^2)\sqrt{1 + x^2}}.$$

Let us use the substitution $x = \tan t$, so that $x = \arctan t$, and $\frac{dx}{1 + x^2} = dt$. This gives the following

$$I_{17} = \int \frac{dt}{\sqrt{1 + \tan^2 t}} = \int \frac{dt}{\sqrt{1/\cos^2 t}} = \int \cos t \, dt = \sin t + C$$

$$= \sin(\arctan x) + C = \frac{x}{\sqrt{1 + x^2}} + C,$$

where $C \in \mathbb{R}$.

$$I_{18} = \int \frac{dx}{\sqrt{x\sqrt{x} - x^2}}$$

**Solution 1.**

For the given integral we can write the following

$$I_{18} = \int \left(x^{3/2} - x^2\right)^{-1/2} dx = \int \left(x^{3/2}\right)^{-1/2} \left(1 - x^{1/2}\right)^{-1/2} dx$$
$$= \int x^{-3/4} \left(1 - x^{1/2}\right)^{-1/2} dx.$$

This is a binomial integral, from the form $\int x^m (ax^n + b)^p dx$, where $m = -3/4, n = 1/2, p = -1/2 \notin \mathbb{Z}$. But $\frac{m+1}{n} + p = 0 \in \mathbb{Z}$. Therefore we can write

$$I_{18} = \int x^{-3/4} \left(x^{1/2}\right)^{-1/2} \left(x^{-1/2} - 1\right)^{-1/2} dx = \int x^{-1} \left(x^{-1/2} - 1\right)^{-1/2} dx.$$

Let us use the substitution $x^{-1/2} - 1 = t^2$, so that $x = \frac{1}{(1+t^2)^2}$, and $dx = -\frac{4t}{(1+t^2)^3} dt$. This gives the following

$$I_{18} = -4 \int (1+t^2)^2 (t^2)^{-1/2} \frac{t}{(1+t^2)^3} dt = -4 \int \frac{dt}{1+t^2} = -4\arctan t + C$$
$$= -4\arctan\left(\sqrt{\frac{1 - \sqrt{x}}{\sqrt{x}}}\right) + C,$$

where $C \in \mathbb{R}$.

**Solution 2.**

Let us use the substitution $\sqrt{x} = t > 0$, so that $x = t^2$, and $dx = 2t\,dt$. This gives the following

$$I_{18} = 2 \int \frac{t}{\sqrt{t^3 - t^4}} dt = 2 \int \frac{t}{\sqrt{t^3 - t^4}} dt = 2 \int \frac{dt}{\sqrt{t - t^2}} = 2 \int \frac{t}{|t|\sqrt{t - t^2}} dt$$
$$= 2 \int \frac{dt}{\sqrt{t - t^2}} = 2 \int \frac{dt}{\sqrt{\frac{1}{4} - \left(t - \frac{1}{2}\right)^2}} = 2\arcsin\left(\frac{t - (1/2)}{1/2}\right) + C$$
$$= 2\arcsin(2t - 1) + C = 2\arcsin(2\sqrt{x} - 1) + C,$$

where $C \in \mathbb{R}$.

**Solution 3.**

Let us use the substitution $\sqrt[4]{x} = t > 0$, so that $x = t^4$, and $dx = 4t^3 dt$. This gives the following

$$I_{18} = 4 \int \frac{t^3}{\sqrt{t^6 - t^8}} dt = 4 \int \frac{t^3}{|t^3|\sqrt{1 - t^2}} dt = \int \frac{dt}{\sqrt{1 - t^2}} = 4 \arcsin t + C$$
$$= 4 \arcsin\left(\sqrt[4]{x}\right) + C,$$

where $C \in \mathbb{R}$.

$$I_{19} = \int \frac{x - 1}{x + x^2 \ln x} dx$$

**Solution.**
For the given integral we can write the following

$$I_{19} = \int \frac{x - 1}{x^2 \left(\frac{1}{x} + \ln x\right)} dx = \int \frac{\frac{1}{x} - \frac{1}{x^2}}{\frac{1}{x} + \ln x} dx = \int \frac{\left(\frac{1}{x} + \ln x\right)'}{\frac{1}{x} + \ln x} dx$$
$$= \ln\left|\frac{1}{x} + \ln x\right| + C,$$

where $C \in \mathbb{R}$.

$$I_{20} = \int \csc x \sec x \, dx$$

**Solution.**
For the given integral we can write the following

$$I_{20} = \int \frac{dx}{\sin x \cos x} = 2 \int \frac{dx}{\sin(2x)}.$$

For the last integral, let us use the substitution $2x = t$, so that $2dx = dt$. This gives the following

$$I_{20} = \int \frac{dt}{\sin t} = \ln\left|\tan\left(\frac{t}{2}\right)\right| + C = \ln|\tan x| + C,$$

where $C \in \mathbb{R}$.

# Chapter 12

# The Solutions to the 2019 MIT Integration Bee, Qualifying Test

$$I_1 = \int_0^{2\pi} \tan(\cos x)dx$$

**Solution 1.**

Let us use the substitution $x + 2\pi = t$, so that $dx = dt$. If $x = 0$, then $t = 2\pi$, and if $x = 2\pi$, then $t = 4\pi$. With these new values, we get the following

$$I_1 = \int_0^{4\pi} \tan(\cos t)dt = \underbrace{\int_0^{2\pi} \tan(\cos t)dt}_{I_1} + \underbrace{\int_{2\pi}^{4\pi} \tan(\cos t)dt}_{:=J}.$$

For the integral $J$, let us use the substitution $t - 2\pi = u$, so that $dt = du$. If $t = 2\pi$, then $u = 0$, and if $t = 4\pi$, then $u = 2\pi$. With these new values, we get the following

$$I_1 = \underbrace{\int_0^{2\pi} \tan(\cos t)dt}_{I_1} + \underbrace{\int_0^{2\pi} \tan(\cos u)du}_{I_1} = 2I_1.$$

Thus, we get

$$I_1 = \int_0^{2\pi} \tan(\cos x)dx = 0.$$

**Solution 2.**

For the given integral we can write the following

$$I_1 = \int_0^\pi \tan(\cos x)dx + \underbrace{\int_\pi^{2\pi} \tan(\cos x)dx}_{:=J}.$$

For the integral $J$, let us use the substitution $x = t + \pi$, so that $dx = dt$. If $x = \pi$, then $t = 0$, and if $x = 2\pi$, then $t = \pi$. With these new values, we get the following

$$I_1 = \int_0^\pi \tan(\cos x)dx + \int_0^\pi \tan(\cos(t + \pi))dt$$
$$= \int_0^\pi \tan(\cos x)dx - \int_0^\pi \tan(\cos t)dt = 0.$$

$$\boxed{I_2 = \int \frac{x+1}{x(x + \ln x)}dx}$$

**Solution.**

For the given integral we can write the following

$$I_2 = \int \frac{1 + \frac{1}{x}}{x + \ln x}dx = \int \frac{(x + \ln x)'}{x + \ln x}dx = \ln|x + \ln x| + C,$$

where $C \in \mathbb{R}$.

$$\boxed{I_3 = \int \left(e^{x+e^x} + e^{x-e^x}\right)dx}$$

**Solution.**

For the given integral we can write the following

$$I_3 = \int \left(e^x e^{e^x} + e^x e^{-e^x}\right)dx = \int e^x \left(e^{e^x} + e^{-e^x}\right)dx.$$

Now, let us use the substitution $e^x = t$, so that $e^x dx = dt$. This gives the following

$$I_3 = \int \left(e^t + e^{-t}\right)dt = e^t - e^{-t} + C = e^{e^x} - e^{-e^x} + C,$$

where $C \in \mathbb{R}$.

$$\boxed{I_4 = \int_{-1/2}^{1/2} \frac{dx}{1 - x^2}}$$

**Solution.**

For the given integral we can write the following

$$I_4 = \left[\frac{1}{2}\ln\left|\frac{1+x}{1-x}\right|\right]_{-1/2}^{1/2} = \frac{1}{2}\left(\ln\left|\frac{1+(1/2)}{1-(1/2)}\right| - \ln\left|\frac{1-(1/2)}{1+(1/2)}\right|\right)$$

$$= \frac{1}{2}\left(\ln 3 - \ln\left(\frac{1}{3}\right)\right) = \ln 3.$$

**Remark.** We can calculate the integral $I_4$ according to the method of partial fractions, where

$$\frac{1}{1-x^2} = \frac{1}{2}\left(\frac{1}{1-x} + \frac{1}{1+x}\right).$$

Then, we get

$$I_4 = \frac{1}{2}\int_{-1/2}^{1/2}\left(\frac{1}{1-x} + \frac{1}{1+x}\right)dx = \frac{1}{2}\left[-\ln|1-x| + \ln|1+x|\right]_{-1/2}^{1/2}$$

$$= \frac{1}{2}\left[\ln\left|\frac{1+x}{1-x}\right|\right]_{-1/2}^{1/2} = \frac{1}{2}\left(\ln 3 - \ln\left(\frac{1}{3}\right)\right) = \ln 3.$$

$$\boxed{I_5 = \int_0^2 2^{\ln x}\,dx}$$

**Solution.**

For the given integral we can write the following

$$I_5 = \lim_{\varepsilon \to 0^+}\int_\varepsilon^2 2^{\ln x}\,dx.$$

Let us use the substitution $\ln x = t$, so that $x = e^t$, and $dx = e^t dt$. If $x = \varepsilon$, then $t = \ln\varepsilon$, and if $x = 2$, then $t = \ln 2$. With these new values, we get the following

$$I_5 = \lim_{\varepsilon \to 0^+}\int_{\ln\varepsilon}^{\ln 2}(2e)^t\,dt = \lim_{\varepsilon \to 0^+}\left[\frac{(2e)^t}{\ln(2e)}\right]_{\ln\varepsilon}^{\ln 2} = \lim_{\varepsilon \to 0^+}\left(\frac{(2e)^{\ln 2}}{\ln(2e)} - \frac{(2e)^{\ln\varepsilon}}{\ln(2e)}\right)$$

$$= \frac{(2e)^{\ln 2}}{\ln(2e)} = \frac{2^{\ln 2}e^{\ln 2}}{\ln(2) + \ln(e)} = \frac{2^{1+\ln 2}}{1 + \ln 2}.$$

$$\boxed{I_6 = \int_{-2\pi}^{2\pi}(\cos(3x) + \sin(2x))\,(-\sin(2019x) + \cos(3x))\,dx}$$

**Solution.**

For the given integral we can write the following

$$
I_6 = - \int_{-2\pi}^{2\pi} \underbrace{\sin(2019x)\cos(3x)\,dx}_{\text{odd function}} + \int_{-2\pi}^{2\pi} \underbrace{\cos^2(3x)\,dx}_{\text{even function}}
$$

$$
- \int_{-2\pi}^{2\pi} \underbrace{\sin(2x)\sin(2019x)\,dx}_{\text{even function}} + \int_{-2\pi}^{2\pi} \underbrace{\sin(2x)\cos(3x)\,dx}_{\text{odd function}}
$$

$$
= 0 + 2\int_{0}^{2\pi} \cos^2(3x)\,dx - 2\int_{0}^{2\pi} \sin(2x)\sin(2019x)\,dx + 0
$$

$$
= 2\int_{0}^{2\pi} \frac{1+\cos(6x)}{2}\,dx - 2\int_{0}^{2\pi} -\frac{1}{2}\left(\cos(2021x) - \cos(-2017x)\right)dx
$$

$$
= \int_{0}^{2\pi} (1+\cos(6x))\,dx + \int_{0}^{2\pi} \left(\cos(2021x) - \cos(2017x)\right)dx
$$

$$
= 2\pi + \left[\frac{1}{6}\sin(6x)\right]_0^{2\pi} + \left[\frac{1}{2021}\sin(2021x)\right]_0^{2\pi} - \left[\frac{1}{2017}\sin(2017x)\right]_0^{2\pi}
$$

$$
= 2\pi.
$$

$$
\boxed{I_7 = \int \cos x \left(\cos(\sin x)\right)\cos\left(\sin(\sin x)\right)dx}
$$

**Solution.**

Let us use the substitution $\sin x = t$, so that $\cos x\,dx = dt$. This gives the following

$$
I_7 = \int \cos t \cos(\sin t)\,dt.
$$

Now, let us use the substitution $\sin t = u$, so that $\cos t\,dt = du$. This gives the following

$$
I_7 = \int \cos u\,du = \sin u + C = \sin(\sin t) + C = \sin(\sin(\sin x)) + C,
$$

where $C \in \mathbb{R}$.

**Remark.**  For integral $I_7$, we can use the substitution $\sin(\sin x) = t$, so that $\cos x \left(\cos(\sin x)\right)dx = dt$. This gives the following

$$
I_7 = \int \cos t\,dt = \sin t + C = \sin(\sin(\sin x)) + C.
$$

$$
\boxed{I_8 = \int_{0}^{\infty} \frac{e^{-2019/(4t^2)}}{t^2}\,dt}
$$

**Solution.**

Let us use the substitution $\frac{\sqrt{2019}}{2t} = u$, so that $\frac{2019}{4t^2} = u^2$ and $\frac{dt}{t^2} = -\frac{2}{\sqrt{2019}}du$. If $t \to 0$, then $u \to \infty$, and if $t \to \infty$, then $u \to 0$. This gives the following

$$I_8 = -\frac{2}{\sqrt{2019}} \int_\infty^0 e^{-u^2} du = \frac{2}{\sqrt{2019}} \int_0^\infty e^{-u^2} du = \frac{1}{\sqrt{2019}} \int_{-\infty}^\infty e^{-u^2} du$$

$$= \sqrt{\frac{\pi}{2019}}.$$

Where we used the Gaussian integral (1.14).

$$\boxed{I_9 = \int \sin\left(\sqrt{x}\right) dx}$$

**Solution.**

Let us use the substitution $\sqrt{x} = t$, so that $x = t^2$, and $dx = 2tdt$. This gives the following

$$I_9 = 2 \int t \sin t dt.$$

Now, by using the integration by parts (assume $u = t$, $dv = \sin t dt$), we get

$$I_9 = 2\left(-t\cos t + \sin t\right) + C = -2\sqrt{x}\cos\left(\sqrt{x}\right) + 2\sin\left(\sqrt{x}\right) + C,$$

where $C \in \mathbb{R}$.

$$\boxed{I_{10} = \int_0^1 \frac{\sqrt{x}}{1+x} dx}$$

**Solution.**

Let us use the substitution $x = t^2$, so that $dx = 2tdt$. If $x = 0$, then $t = 0$, and if $x = 1$, then $t = 1$. This gives the following

$$I_{10} = 2 \int_0^1 \frac{t^2}{1+t^2} dt = 2 \int_0^1 \left(1 - \frac{1}{1+t^2}\right) dt = 2\left[t - \arctan t\right]_0^1$$

$$= 2\left(1 - \frac{\pi}{4}\right) = 2 - \frac{\pi}{2}.$$

$$\boxed{I_{11} = \int_0^{2\pi} \cos x \cos(2x) \cos(3x) dx}$$

**Solution.**

For the given integral we can write the following

$$I_{11} = \frac{1}{2} \int_0^{2\pi} (\cos(3x) + \cos x) \cos(3x) dx$$

$$= \frac{1}{2} \int_0^{2\pi} \left( \cos^2(3x) + \cos x \cos(3x) \right) dx$$

$$= \frac{1}{2} \int_0^{2\pi} \left( \frac{1 + \cos(6x)}{2} + \frac{1}{2} (\cos(4x) + \cos(2x)) \right) dx$$

$$= \frac{1}{2} \int_0^{2\pi} \left( \frac{1}{2} + \frac{1}{2} \cos(6x) + \frac{1}{2} \cos(4x) + \frac{1}{2} \cos(2x) \right) dx$$

$$= \frac{1}{2} \int_0^{2\pi} \frac{1}{2} dx = \frac{1}{4} (2\pi) = \frac{\pi}{2}.$$

$$\boxed{\lim_{n \to \infty} \int_{-\infty}^{\infty} e^{-x^{2n}} dx}$$

**Solution.**

At the first, we note that the integrand $f(x) := e^{-x^{2n}}$ is an even function on $\mathbb{R}$, since $f(-x) = -f(x), \forall x \in \mathbb{R}$. Thus, we can write the following

$$L := \lim_{n \to \infty} \int_{-\infty}^{\infty} e^{-x^{2n}} dx = 2 \lim_{n \to \infty} \int_0^{\infty} e^{-x^{2n}} dx.$$

We note that, when $x \in [0, 1[$, we have

$$\lim_{n \to \infty} -x^{2n} = 0 \quad \Longrightarrow \quad \lim_{n \to \infty} e^{-x^{2n}} = 1,$$

and, when $x \in ]1, \infty[$, we have

$$\lim_{n \to \infty} -x^{2n} = -\infty \quad \Longrightarrow \quad \lim_{n \to \infty} e^{-x^{2n}} = 0.$$

Thus for the given limit, we have

$$L = 2 \lim_{n \to \infty} \left( \int_0^1 e^{-x^{2n}} dx + \int_1^{\infty} e^{-x^{2n}} dx \right).$$

Now, Let us denote

$$f_1 : [0, 1[ \longrightarrow \mathbb{R}, \quad f_1(x) = e^{-x^{2n}}, \quad \forall x \in [0, 1[,$$

$$f_2 : ]1, \infty[ \longrightarrow \mathbb{R}, \quad f_2(x) = e^{-x^{2n}}, \quad \forall x \in ]1, \infty[.$$

We note that

$$f_1(x) = e^{-x^{2n}} = \frac{1}{e^{x^{2n}}} \leqslant 1 \quad \forall x \in [0,1[ \text{ and } \forall n \geqslant 1,$$

and

$$f_2(x) = e^{-x^{2n}} = \frac{1}{e^{x^{2n}}} \leqslant \frac{1}{e^x} \quad \forall x \in ]1, \infty[ \text{ and } \forall n \geqslant 1.$$

The functions $g_1(x) := 1$ and $g_2(x) := e^{-x}$ are integrable on $[0,1[$ and $]1, \infty[$ respectively. Where

$$\int_1^\infty e^{-x} dx = \lim_{a \to \infty} \int_1^a e^{-x} dx = \lim_{a \to \infty} \left( \frac{1}{e} - \frac{1}{e^a} \right) = \frac{1}{e}.$$

Thus, by Lebesgue's dominated convergence theorem (see Subsec. 1.9.4), we can write

$$L = 2 \left( \int_0^1 \lim_{n \to \infty} e^{-x^{2n}} dx + \int_1^\infty \lim_{n \to \infty} e^{-x^{2n}} dx \right)$$

$$= 2 \left( \int_0^1 dx + \int_1^\infty 0 dx \right) = 2(1 + 0) = 2.$$

$$\boxed{I_{13} = \int_0^e x^{1/\ln x} dx}$$

**Solution 1.**

For the integrand of $I_{13}$, because $a^b = e^{b \ln a}$, $\forall a > 0, b \in \mathbb{R}$, we can write the following

$$x^{1/\ln x} = e^{\ln x (1/\ln x)} = e^1 = e.$$

Therefore, we get

$$I_{13} = \int_0^e e\, dx = [ex]_0^e = e^2 - 0 = e^2.$$

**Solution 2.**

Let we calculate the indefinite integral

$$I := \int x^{1/\ln x} dx.$$

For this, let us use the substitution $\frac{1}{\ln x} = t$, so that $x = e^{1/t}$, and $dx = -\frac{e^{1/t}}{t^2} dt$. This gives the following

$$I_{13} = \int \left( e^{1/t} \right)^t \left( -\frac{e^{1/t}}{t^2} \right) dt = e \int e^{1/t} \left( -\frac{1}{t^2} \right) dt.$$

Now, for the last integral, let us use the substitution $1/t = u$, so that $-dt/t^2 = du$. This gives the following

$$I_{13} = e \int e^u du = ee^u + C = ee^{1/t} + C = ee^{\ln x} + C = ex + C,$$

where $C \in \mathbb{R}$. Therefore, for the given integral $I_{13}$, we get

$$I_{13} = \lim_{\varepsilon \to 0} \int_\varepsilon^e x^{1/\ln x} dx = \lim_{\varepsilon \to 0} [ex]_\varepsilon^e = \lim_{\varepsilon \to 0} \left(e^2 - e\varepsilon\right) = e^2.$$

$$I_{14} = \int_0^{\pi/100} \frac{\sin(20x) + \sin(19x)}{\cos(20x) + \cos(19x)} dx$$

**Solution.**

For the given integral we can write the following

$$I_{14} = \int_0^{\pi/100} \frac{2\sin(39x/2)\cos(x/2)}{2\cos(39x/2)\cos(x/2)} dx$$

$$= -\frac{2}{39} \int_0^{\pi/100} \frac{(-39/2)\sin(39x/2)}{\cos(39x/2)} dx$$

$$= -\frac{2}{39} \left[\ln\left|\cos\left(\frac{39x}{2}\right)\right|\right]_0^{\pi/100} = -\frac{2}{39}\ln\left(\cos\left(\frac{39\pi}{200}\right)\right).$$

$$I_{15} = \int \left(e^x \cos^2 x + e^x \sin x \cos x - e^x \sin^2 x\right) dx$$

**Solution.**

For the given integral we can write the following

$$I_{15} = \int e^x (\cos^2 x - \sin^2 x) dx + \int e^x \sin x \cos x \, dx$$

$$= \int e^x \cos(2x) dx + \frac{1}{2} \int e^x \sin(2x) dx.$$

But, for $a, b \in \mathbb{R}$ such that $a^2 + b^2 \neq 0$, we have (see (1.1) and (1.2))

$$\int e^{ax} \cos(bx) dx = \frac{e^{ax}}{a^2 + b^2} \left(a\cos(bx) + b\sin(bx)\right) + C,$$

and

$$\int e^{ax} \sin(bx) dx = \frac{e^{ax}}{a^2 + b^2} \left(a\sin(bx) - b\cos(bx)\right) + C,$$

where $C \in \mathbb{R}$. Therefore, we get

$$I_{15} = \frac{e^x}{1+4}\left(\cos(2x) + 2\sin(2x)\right) + \frac{1}{2}\frac{e^x}{1+4}\left(\sin(2x) - 2\cos(2x)\right) + C$$

$$= \frac{e^x}{5}\left(\cos(2x) + 2\sin(2x) + \frac{1}{2}\sin(2x) - \cos(2x)\right) + C$$

$$= \frac{e^x}{2}\sin(2x) + C.$$

$$\boxed{I_{16} = \int_0^{\pi/2} \frac{\sin x}{\sin\left(x + \frac{\pi}{4}\right)}\,dx}$$

**Solution.**

Let us use the substitution $x + \frac{\pi}{4} = t$, so that $dx = dt$. If $x = 0$, then $t = \pi/4$, and if $x = \pi/2$, then $t = 3\pi/4$. With these new values, we get the following

$$I_{16} = \int_{\pi/4}^{3\pi/4} \frac{\sin\left(t - \frac{\pi}{4}\right)}{\sin t}\,dt = \int_{\pi/4}^{3\pi/4} \frac{\sin t \cos\left(\pi/4\right) - \cos t \sin\left(\pi/4\right)}{\sin t}\,dt$$

$$= \frac{1}{\sqrt{2}}\int_{\pi/4}^{3\pi/4}\left(1 - \frac{\cos t}{\sin t}\right)dt = \frac{1}{\sqrt{2}}\left[t - \ln|\sin t|\right]_{\pi/4}^{3\pi/4}$$

$$= \frac{1}{\sqrt{2}}\left(\frac{3\pi}{4} - \ln\left|\sin\left(\frac{3\pi}{4}\right)\right| - \frac{\pi}{4} + \ln\left|\sin\left(\frac{\pi}{4}\right)\right|\right) = \frac{\pi}{2\sqrt{2}}.$$

$$\boxed{I_{17} = \int \frac{dx}{x + \sqrt[3]{x}}}$$

**Solution.**

Let us use the substitution $x = t^3$, so that $dx = 3t^2 dt$. This gives the following

$$I_{17} = 3\int \frac{t^2}{t^3 + t}\,dt = \frac{3}{2}\int \frac{2t}{1+t^2}\,dt = \frac{3}{2}\ln(1+t^2) + C$$

$$= \frac{3}{2}\ln\left(1 + x^{2/3}\right) + C,$$

where $C \in \mathbb{R}$.

$$\boxed{I_{18} = \int_0^2 x^{x^2+1}\left(2\ln x + 1\right)dx}$$

**Solution.**

Let us calculate the indefinite integral

$$I := \int x^{x^2+1}\left(2\ln x + 1\right)dx.$$

For this integral, we can write

$$I = \int x x^{x^2} \left( 2 \ln x + 1 \right) dx = \int e^{x^2 \ln x} \left( x + 2x \ln x \right) dx.$$

Let us use the substitution $x^2 \ln x = t$, so that $(x + 2x \ln x)dx = dt$. This gives the following

$$I = \int e^t dt = e^t + C = e^{x^2 \ln x} + C,$$

where $C \in \mathbb{R}$. Now, for the given integral, we get

$$I_{18} = \lim_{\varepsilon \to 0^+} \int_{\varepsilon}^{2} x^{x^2+1} \left( 2 \ln x + 1 \right) dx = \lim_{\varepsilon \to 0^+} \left[ e^{x^2 \ln x} \right]_{\varepsilon}^{2} = e^{4 \ln 2} - \lim_{\varepsilon \to 0^+} e^{\varepsilon^2 \ln \varepsilon}$$

$$= e^{\ln(2^4)} - e^0 = 15.$$

$$\boxed{I_{19} = \int \frac{2x^3 - 1}{x(x^3 + 1)} dx}$$

**Solution.**

Let us use the substitution $x^3 = t$, so that $x = t^{1/3}$, and $dx = \frac{dt}{3t^{2/3}}$. This gives the following

$$I_{19} = \int \frac{2t - 1}{t^{1/3}(t + 1)} \frac{dt}{3t^{2/3}} = \frac{1}{3} \int \frac{2t - 1}{t(t + 1)} dt = \frac{1}{3} \int \left( -\frac{1}{t} + \frac{3}{t + 1} \right) dt$$

$$= \frac{1}{3} \left( -\ln |t| + 3 \ln |t + 1| \right) + C = -\frac{1}{3} \ln |x^3| + \ln |x^3 + 1| + C$$

$$= -\ln |x| + \ln |x^3 + 1| + C = \ln \left| x^2 + \frac{1}{x} \right| + C,$$

where $C \in \mathbb{R}$.

**Remark.**  We can calculate the integral $I_{19}$ according to the method of partial fractions (but for this method, more time will be required.), where

$$\frac{2x^3 - 1}{x(x^3 + 1)} = \frac{2x^3 - 1}{x(x + 1)(x^2 - x + 1)} = -\frac{1}{x} + \frac{1}{x + 1} + \frac{2x - 1}{x^2 - x + 1}.$$

$$\boxed{I_{20} = \int \cos \left( \arctan x \right) dx}$$

**Solution.**

Let us use the substitution $\arctan x = t$, and find a formula to the integrand $f(x) :=$ $\cos(\arctan x)$ by $x$, such that we can calculate $I_{20}$ in an easy way.

From the substitution $t = \arctan x$, we get $\tan t = x$ and thus (see the inserted figure below, and the Subsec. 1.5.1)

$$\cos(\arctan x) = \cos t = \frac{1}{\sqrt{1+x^2}}.$$

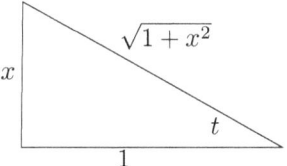

Therefore, for the given integral $I_{20}$, we have

$$I_{20} = \int \cos\left(\arctan x\right) dx = \int \frac{dx}{\sqrt{1+x^2}} = \ln\left|x + \sqrt{1+x^2}\right| + C,$$

where $C \in \mathbb{R}$.

**Remark.** From the previous figure, we can also find $\sin(\arctan x) = \frac{x}{\sqrt{1+x^2}}$. Thus, we can calculate the following integral

$$\int \sin(\arctan x) dx = \int \frac{x}{\sqrt{1+x^2}} dx = \frac{1}{2}\sqrt{1+x^2} + C.$$

Similarly, we can calculate

$$I = \int \tan(\arccos x) dx. \qquad (12.1)$$

By using the substitution $t = \arccos x$ with $x \in ]-\infty, -1] \cup [1, \infty[$, we find $\cos t = x$, and $\tan(\arccos x) = \tan t = \frac{\sqrt{1+x^2}}{x}$ (see the following figure, and the Subsec. 1.5.1)

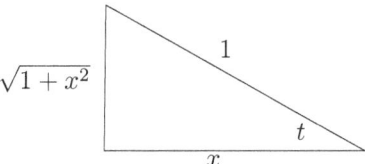

Thus for the integral (12.1), we get

$$I = \int \frac{\sqrt{1+x^2}}{x} dx.$$

Now, let us use the substitution $\sqrt{1 + x^2} = u$, so that $x^2 = u^2 - 1$, and $dx = \frac{\sqrt{1+x^2}}{x} du = \frac{u}{x} du$. This gives the following

$$I = \int \frac{u}{x} \frac{u}{x} du = \int \frac{u^2}{u^2 - 1} du = \int \left( 1 + \frac{1}{u^2 - 1} \right) du = \int \left( 1 - \frac{1}{1 - u^2} \right) du$$

$$= u - \frac{1}{2} \ln \left| \frac{1 + u}{1 - u} \right| + C = \sqrt{1 + x^2} - \frac{1}{2} \ln \left| \frac{1 + \sqrt{1 + x^2}}{1 - \sqrt{1 + x^2}} \right| + C,$$

where $C \in \mathbb{R}$.

# Chapter 13

# The Solutions to the 2020 MIT Integration Bee, Qualifying Test

$$I_1 = \int \frac{\ln(2x)}{x \ln x} dx$$

**Solution.**

For the given integral we can write the following

$$I_1 = \int \frac{\ln x + \ln 2}{x \ln x} dx.$$

Let us use the substitution $\ln x = t$, so that $dx/x = dt$. This gives the following

$$I_1 = \int \frac{t + \ln 2}{t} dt = \int \left(1 + \frac{\ln 2}{t}\right) dt = t + \ln 2 \ln |t| + C$$

$$= \ln x + \ln 2 \ln |\ln x| + C,$$

where $C \in \mathbb{R}$.

$$I_2 = \int_0^\infty \frac{dx}{e^x + 1}$$

**Solution.**

For the given integral we can write the following

$$I_2 = \lim_{a \to \infty} \int_0^a \frac{dx}{e^x + 1} = \lim_{a \to \infty} \int_0^a \frac{e^{-x}}{1 + e^{-x}} dx = -\lim_{a \to \infty} \int_0^a \frac{-e^{-x}}{1 + e^{-x}} dx$$
$$= -\lim_{a \to \infty} \left[ \ln\left(1 + e^{-x}\right) \right]_0^a = -\lim_{a \to \infty} \left( \ln\left(1 + e^{-a}\right) - \ln 2 \right)$$
$$= -(0 - \ln 2) = \ln 2.$$

$$\boxed{I_3 = \int_e^{e^e} \frac{\ln x \ln (\ln x)}{x} dx}$$

**Solution.**

Let us use the substitution $\ln x = t$, so that $dx/x = dt$. If $x = e$, then $t = 1$, and if $x = e^e$, then $t = e$. With these new values, we get the following

$$I_3 = \int_1^e t \ln t \, dt.$$

Now, by using the integration by parts, let us assume

$$u = \ln t \implies du = \frac{dt}{t}, \quad dv = t dt \implies v = \frac{t^2}{2}.$$

Thus, we have

$$I_3 = \left[ \frac{1}{2} t^2 \ln t \right]_1^e - \frac{1}{2} \int_1^e t \, dt = \frac{e^2}{2} - \left[ \frac{t^2}{4} \right]_1^e = \frac{e^2}{2} - \left( \frac{e^2}{4} - \frac{1}{4} \right) = \frac{e^2 + 1}{4}.$$

$$\boxed{I_4 = \int_0^1 \ln \left( \frac{1 + x}{1 - x} \right) dx}$$

**Solution.**

For the given integral we can write the following

$$I_4 = \lim_{a \to 1^-} \int_0^a \ln \left( \frac{1 + x}{1 - x} \right) dx.$$

By using the integration by parts, let us assume

$$u = \ln \left( \frac{1 + x}{1 - x} \right) \implies du = \left( \frac{1}{1 + x} - \frac{1}{1 - x} \right) dx, \quad dv = dx \implies v = x.$$

Thus, we have

$$I_4 = \lim_{a \to 1^-} \left( \left[ x \ln \left( \frac{1+x}{1-x} \right) \right]_0^a - \int_0^a \left( \frac{x}{1+x} + \frac{x}{1-x} \right) dx \right)$$

$$= \lim_{a \to 1^-} \left( a \ln \left( \frac{1+a}{1-a} \right) + \int_0^a \left( \frac{1}{1+x} - \frac{1}{1-x} \right) dx \right)$$

$$= \lim_{a \to 1^-} \left( a \ln(1+a) - a \ln(1-a) + [\ln|1+x| + \ln|1-x|]_0^a \right)$$

$$= \lim_{a \to 1^-} \left( a \ln(1+a) - a \ln(1-a) + \ln(1+a) + \ln(1-a) \right)$$

$$= \lim_{a \to 1^-} \left( (1+a) \ln(1+a) + (1-a) \ln(1-a) \right)$$

$$= 2 \ln 2 + \lim_{a \to 1^-} (1-a) \ln(1-a) = 2 \ln 2 + \lim_{b \to 0^+} b \ln b = 2 \ln 2.$$

$$\boxed{I_5 = \int \frac{dx}{x^2 + (x-1)^2}}$$

**Solution.**

For the given integral we can write the following

$$I_5 = \int \frac{dx}{2x^2 - 2x + 1} = \frac{1}{2} \int \frac{dx}{x^2 - x + \frac{1}{2}} = \frac{1}{2} \int \frac{dx}{x^2 - x + \frac{1}{4} + \frac{1}{4}}$$

$$= \frac{1}{2} \int \frac{dx}{\left( x - \frac{1}{2} \right)^2 + \frac{1}{4}} = \frac{1}{2} \frac{1}{1/2} \arctan \left( \frac{x - \frac{1}{2}}{1/2} \right) + C$$

$$= \arctan(2x - 1) + C,$$

where $C \in \mathbb{R}$.

$$\boxed{I_6 = \int \sqrt{x \sqrt{x \sqrt{x \sqrt{x \ldots}}}}\, dx}$$

**Solution 1.**

Let we set

$$f(x) := \sqrt{x \sqrt{x \sqrt{x \sqrt{x \ldots}}}} > 0.$$

Then

$$f^2(x) = x \sqrt{x \sqrt{x \sqrt{x \ldots}}} \implies f^2(x) = x f(x) \implies f(x)(f(x) - x) = 0.$$

Thus, we have $f(x) = x$ (where $f(x) > 0$). Therefore, for the given integral we get

$$I_6 = \int x\, dx = \frac{x^2}{2} + C,$$

where $C \in \mathbb{R}$.

**Solution 2.**

For the given integral we can write the following

$$
I_6 = \int x^{\frac{1}{2}} \left( \sqrt{x\sqrt{x\sqrt{x\ldots}}} \right)^{\frac{1}{2}} dx = \int x^{\frac{1}{2}} x^{\frac{1}{2}\frac{1}{2}} \left( \sqrt{x\sqrt{x\ldots}} \right)^{\frac{1}{2}\frac{1}{2}} dx
$$

$$
= \int x^{\frac{1}{2}} x^{\frac{1}{2}\frac{1}{2}} x^{\frac{1}{2}\frac{1}{2}\frac{1}{2}} \left( \sqrt{x\ldots} \right)^{\frac{1}{2}\frac{1}{2}\frac{1}{2}} dx = \int x^{\frac{1}{2}} x^{\frac{1}{2}\frac{1}{2}} x^{\frac{1}{2}\frac{1}{2}\frac{1}{2}} x^{\frac{1}{2}\frac{1}{2}\frac{1}{2}\frac{1}{2}} \cdots dx
$$

$$
= \int x^{1/2} x^{(1/2)^2} x^{(1/2)^3} x^{(1/2)^4} \cdots dx = \int x^{\sum_{n=1}^{\infty}(1/2)^n} dx
$$

$$
= \int x^{\frac{1/2}{1-(1/2)}} dx = \int x dx = \frac{x^2}{2} + C,
$$

where $C \in \mathbb{R}$.

$$
\boxed{I_7 = \int \sin^4 x \cos^4 x (\cos x + \sin x)(\cos x - \sin x) dx}
$$

**Solution 1.**

For the given integral we can write the following

$$
I_7 = \int (\sin x \cos x)^4 \left( \cos^2 x - \sin^2 x \right) dx = \int \left( \frac{1}{2} \sin(2x) \right)^4 \cos(2x) dx
$$

$$
= \frac{1}{2^4} \int \sin^4(2x) \cos(2x) dx = \frac{1}{2^4} \frac{1}{2} \int \sin^4(2x)(2\cos(2x)) dx
$$

$$
= \frac{1}{2^5} \left( \frac{1}{5} \sin^5(2x) \right) + C = \frac{\sin^5(2x)}{5(2^5)} + C = \frac{2^5 \sin^5 x \cos^5 x}{5(2^5)} + C
$$

$$
= \frac{1}{5} \sin^5 x \cos^5 x + C,
$$

where $C \in \mathbb{R}$.

**Solution 2.**

For the given integral we can write the following

$$
I_7 = \int (\sin x \cos x)^4 \left( \cos^2 x - \sin^2 x \right) dx.
$$

Let us use the substitution $\sin x \cos x = t$, so that $(\cos^2 x - \sin^2 x) dx = dt$. This gives the following

$$
I_7 = \int t^4 dt = \frac{1}{5} t^5 + C = \frac{1}{5} \sin^5 x \cos^5 x + C,
$$

where $C \in \mathbb{R}$.

$$I_8 = \int \ln(1 + x^2)dx$$

**Solution.**

By using the integration by parts, let us assume

$$u = \ln(1 + x^2) \Longrightarrow du = \frac{2x}{1 + x^2}dx, \quad dv = dx \Longrightarrow v = x.$$

Thus, we have

$$I_8 = x \ln(1 + x^2) - 2 \int \frac{x^2}{1 + x^2}dx$$

$$= x \ln(1 + x^2) - 2 \int \left(1 - \frac{1}{1 + x^2}\right) dx$$

$$= x \ln(1 + x^2) - 2x + 2 \arctan x + C,$$

where $C \in \mathbb{R}$.

$$I_9 = \int_0^{2\pi} (\cos x)^{2020}dx$$

**Solution.**

Let us find a closed general formula for the integral

$$I_n = \int_0^{2\pi} \cos^n xdx$$

by $n$, when $n$ is an even positive integer. For this, we can write

$$I_n = \int_0^{2\pi} \cos^{n-1} x \cos xdx.$$

By using the integration by parts, let us assume

$$u = \cos^{n-1} x \Longrightarrow du = -(n - 1) \cos^{n-2} x \sin xdx,$$

$$dv = \cos xdx \Longrightarrow v = \sin x.$$

Thus, we have

$$I_n = \left[\sin x \cos^{n-1} x\right]_0^{2\pi} + (n-1) \int_0^{2\pi} \cos^{n-2} x \sin^2 x \, dx$$

$$= (n-1) \int_0^{2\pi} \cos^{n-2} x (1 - \cos^2 x) \, dx$$

$$= (n-1) \left( \underbrace{\int_0^{2\pi} \cos^{n-2} x \, dx}_{I_{n-2}} - \underbrace{\int_0^{2\pi} \cos^n x \, dx}_{I_n} \right)$$

$$= (n-1) I_{n-2} - (n-1) I_n.$$

Thus, for $n \geqslant 2$ (where $n$ is an even), we have

$$I_n = \frac{n-1}{n} I_{n-2} = \frac{n-1}{n} \frac{n-3}{n-2} I_{n-4} = \frac{n-1}{n} \frac{n-3}{n-2} \frac{n-5}{n-4} \frac{n-7}{n-6} \cdots \frac{3}{4} \frac{1}{2} I_0.$$

But $I_0 = \int_0^{2\pi} dx = 2\pi$. Therefore, we get

$$I_n = \frac{(n-1)!!}{n!!} 2\pi, \quad \forall n \geqslant 2 \ (n \text{ even}).$$

But

$$(n-1)!! = \frac{n!}{n!!}, \quad \text{and} \quad n!! = 2^{n/2} (n/2)!.$$

Therefore,

$$I_n = 2\pi \frac{n!}{(n!!)^2} = 2\pi \frac{n!}{\left(2^{n/2} (n/2)!\right)^2} = \frac{2\pi n!}{((n/2)!)^2 \, 2^n} = 2^{1-n} \pi \frac{n!}{(n/2)! \, (n/2)!}$$

$$= 2^{1-n} \pi \frac{n!}{\left(\frac{n}{2}\right)! \left(n - \frac{n}{2}\right)!} = 2^{1-n} \pi \binom{n}{n/2} = \frac{\pi}{2^{n-1}} \binom{n}{n/2}.$$

As a result, for the given integral $I_9$, we have

$$I_9 = \int_0^{2\pi} \cos^{2020} x \, dx = \frac{\pi}{2^{2019}} \binom{2020}{1010}.$$

$$\boxed{I_{10} = \int \frac{2x+1}{2x^2 + 2x + 1} \, dx}$$

**Solution.**

For the given integral we can write the following

$$I_{10} = \frac{1}{2} \int \frac{4x+2}{2x^2 + 2x + 1} \, dx = \frac{1}{2} \ln \left| 2x^2 + 2x + 1 \right| + C,$$

where $C \in \mathbb{R}$.

$$I_{11} = \int_{1/\sqrt{2}}^{1} \frac{\arcsin x}{x^3} dx$$

## Solution 1.

Let us calculate the indefinite integral

$$I := \int \frac{\arcsin x}{x^3} dx. \tag{13.1}$$

By using the integration by parts, let us assume

$$u = \arcsin x \Longrightarrow du = \frac{dx}{\sqrt{1-x^2}}, \quad dv = \frac{dx}{x^3} \Longrightarrow v = -\frac{1}{2x^2}.$$

Thus, we have

$$I = -\frac{\arcsin x}{2x^2} + \frac{1}{2} \underbrace{\int \frac{dx}{x^2\sqrt{1-x^2}}}_{:=J}.$$

Now, for the integral $J$, Let us assume that $x > 0$ and use the substitution $1/x = t$, so that $dx = -dt/t^2$. This gives the following

$$J = -\int \frac{1/t^2}{\frac{1}{t^2}\sqrt{1-\frac{1}{t^2}}} dt = -\int \frac{t}{\sqrt{t^2-1}} dt = -\frac{1}{2} \int 2t(t^2-1)^{-1/2} dt$$

$$= -\sqrt{t^2-1} + C_1 = -\sqrt{\frac{1}{x^2}-1} + C_1 = -\frac{\sqrt{1-x^2}}{x} + C_1,$$

where $C_1 \in \mathbb{R}$. Thus for the integral $I$, we have

$$I = -\frac{\arcsin x}{2x^2} - \frac{\sqrt{1-x^2}}{2x} + C,$$

where $C = C_1/2 \in \mathbb{R}$.

Therefore, for the given integral $I_{11}$, we have

$$I_{11} = \left[ -\frac{\arcsin x}{2x^2} - \frac{\sqrt{1-x^2}}{2x} \right]_{1/\sqrt{2}}^{1}$$

$$= -\frac{\arcsin(1)}{2} - \left( \arcsin\left(\frac{1}{\sqrt{2}}\right) - \frac{1/\sqrt{2}}{\sqrt{2}} \right)$$

$$= -\frac{\pi}{4} - \left( -\frac{\pi}{4} - \frac{1}{2} \right) = \frac{1}{2}.$$

**Solution 2.**

For the indefinite integral (13.1), let us use the substitution $\arcsin x = t$, so that $x = \sin t$, and $dx = \cos t\, dt$. This gives the following

$$I = \int t\, \frac{\cos t}{\sin^3 t}\, dt.$$

Now, by using the integration by parts, let us assume

$$u = t \Longrightarrow du = dt, \quad dv = \frac{\cos t}{\sin^3 t}\, dt \Longrightarrow v = -\frac{1}{2\sin^2 t}.$$

Thus, we have

$$I = -\frac{t}{2\sin^2 t} + \frac{1}{2}\int \frac{dt}{\sin^2 t} = -\frac{t}{2\sin^2 t} - \frac{1}{2}\cot t + C$$

$$= -\frac{\arcsin x}{2\sin^2(\arcsin x)} - \frac{1}{2}\cot(\arcsin x) = -\frac{\arcsin x}{2x^2} - \frac{1}{2}\frac{\cos(\arcsin x)}{\sin(\arcsin x)} + C$$

$$= -\frac{\arcsin x}{2x^2} - \frac{\sqrt{1-x^2}}{2x} + C,$$

where $C \in \mathbb{R}$.

Therefore, for the given integral $I_{11}$, we have

$$I_{11} = \left[ -\frac{\arcsin x}{2x^2} - \frac{\sqrt{1-x^2}}{2x} \right]_{1/\sqrt{2}}^{1}$$

$$= -\frac{\arcsin(1)}{2} - \left( \arcsin\left(\frac{1}{\sqrt{2}}\right) - \frac{1/\sqrt{2}}{\sqrt{2}} \right)$$

$$= -\frac{\pi}{4} - \left( -\frac{\pi}{4} - \frac{1}{2} \right) = \frac{1}{2}.$$

$$\boxed{I_{12} = \int_0^{\pi/2} \sin(2x)\cos(\cos x)\, dx}$$

**Solution.**

For the given integral we can write the following

$$I_{12} = 2\int_0^{\pi/2} \sin x \cos x \cos(\cos x)\, dx.$$

Now, let us use the substitution $\cos x = t$, so that $\sin x\, dx = -dt$. If $x = 0$, then $t = 1$, and if $x = \pi/2$, then $t = 0$. With these new values, we get the following

$$I_{12} = -2 \int_1^0 t \cos t\, dt = 2 \int_0^1 t \cos t\, dt.$$

For the last integral, by using the integration by parts (here we assume $u = t, dv = \cos t\, dt$), we have

$$I_{12} = 2 \left( [t \cos t]_0^1 - \int_0^1 \sin t\, dt \right) = 2 \left( \sin(1) + [\cos t]_0^1 \right)$$
$$= 2 \left( \sin(1) + \cos(1) - 1 \right).$$

$$\boxed{I_{13} = \int_0^{2\pi} \sin\left(\sin x - x\right) dx}$$

**Solution 1.**

Let us use the substitution $2\pi - x = t$, so that $dx = -dt$. If $x = 0$, then $t = 2\pi$, and if $x = 2\pi$, then $t = 0$. With these new values, we get the following

$$I_{13} = -\int_{2\pi}^0 \sin\left(\sin(2\pi - t) - 2\pi + t\right) dt = \int_0^{2\pi} \sin\left(-\sin t - 2\pi + t\right) dt$$
$$= \int_0^{\pi} \sin\left(-(\sin t + 2\pi - t)\right) dt = -\int_0^{2\pi} \sin\left(\sin t - t\right) dt = -I_{13}$$

Therefore,

$$2I_{13} = 0 \quad \Longrightarrow \quad I_{13} = 0.$$

**Solution 2.**

By using the King property of integration (1.13), we find

$$I_{13} = \int_0^{2\pi} \sin\left(\sin(2\pi - x) - (2\pi - x)\right) dx = \int_0^{2\pi} \sin\left(\sin(-x) + x\right) dx$$
$$= \int_0^{2\pi} \sin\left(-\sin x + x\right) dx = -\underbrace{\int_0^{2\pi} \sin\left(\sin x - x\right) dx}_{I_{13}}.$$

Therefore,

$$2I_{13} = 0 \quad \Longrightarrow \quad I_{13} = 0.$$

**Solution 3.**

Let us use the substitution $x - \pi = t$, so that $dx = dt$. If $x = 0$, then $t = -\pi$, and if $x = 2\pi$, then $t = \pi$. With these new values, we get the following

$$I_{13} = \int_\pi^\pi \sin\left(\sin(\pi + t) - \pi - t\right) dt = \int_{-\pi}^\pi \sin\left(-\sin t - \pi - t\right) dt$$

$$= \int_{-\pi}^\pi \sin\left(-\left(\sin t + \pi + t\right)\right) dt = -\int_{-\pi}^{2\pi} \sin\left(\pi + \sin t + t\right) dt$$

$$= \int_{-\pi}^\pi \sin(t + \sin t) dt.$$

But the function $\sin(t + \sin t)$ is an odd function on $[-\pi, \pi]$, Thus $I_{13} = 0$.

$$\boxed{I_{14} = \int \left(\frac{1}{x - 1} + \frac{\sum_{k=0}^{2018}(k + 1)x^k}{\sum_{k=0}^{2019} x^k}\right) dx}$$

**Solution.**

For the given integral we can write the following

$$I_{14} = \int \frac{dx}{x - 1} + \underbrace{\int \frac{\sum_{k=0}^{2018}(k + 1)x^k}{\sum_{k=0}^{2019} x^k} dx}_{:=J}.$$

For the integral $J$, let us use the substitution $\sum_{k=0}^{2019} x^k = t$, so that

$$dt = \left(\sum_{k=0}^{2019} kx^{k-1}\right) dx = \left(\sum_{k=1}^{2019} kx^{k-1}\right) dx = \left(\sum_{k=0}^{2018}(k + 1)x^k\right) dx.$$

This gives the following

$$J = \int \frac{dt}{t} = \ln|t| + C = \ln\left|\sum_{k=0}^{2019} x^k\right| + C,$$

where $C \in \mathbb{R}$.

Therefore, for the given integral $I_{14}$, we get

$$I_{14} = \ln|x - 1| + \ln\left|\sum_{k=0}^{2019} x^k\right| + C = \ln\left|(x - 1)\sum_{k=0}^{2019} x^k\right| + C$$

$$= \ln\left|(x - 1)\left(1 + x + x^2 + \ldots + x^{2019}\right)\right| + C$$

$$= \ln\left|x + x^2 + \ldots + x^{2019} + x^{2020} - 1 - x - x^2 - \ldots - x^{2019}\right| + C$$

$$= \ln\left|1 - x^{2020}\right| + C,$$

where $C \in \mathbb{R}$.

$$I_{15} = \int_0^{\pi/2} \frac{dx}{\tan^{\sqrt{2020}} x + 1} \qquad (13.2)$$

**Solution.**

Let us use the substitution $\frac{\pi}{2} - x = t$, so that $dx = -dt$. If $x = 0$, then $t = \pi/2$, and if $x = \pi/2$, then $t = 0$. With these new values, we get the following

$$I_{15} = -\int_{\pi/2}^0 \frac{dt}{1 + (\cot t)^{\sqrt{2020}}} = \int_0^{\pi/2} \frac{(\tan t)^{\sqrt{2020}}}{1 + (\tan t)^{\sqrt{2020}}} dt. \qquad (13.3)$$

By summing (13.2) and (13.3), we get

$$2I_{15} = \int_0^{\pi/2} \frac{1 + (\tan x)^{\sqrt{2020}}}{1 + (\tan x)^{\sqrt{2020}}} dx = \int_0^{\pi/2} dx = \frac{\pi}{2}.$$

Therefore

$$I_{15} = \frac{\pi}{4}.$$

**Remark.** By using substitution $\frac{\pi}{2} - x = t$, we find that

$$I = \int_0^{\pi/2} \frac{\cos^n x}{\sin^n x + \cos^n x} dx = \frac{\pi}{4}, \qquad \forall n \geqslant 0. \qquad (13.4)$$

Indeed, when $\frac{\pi}{2} - x = t$, so that $dx = -dt$. If $x = 0$, then $t = \pi/2$, and if $x = \pi/2$, then $t = 0$. Then we get the following

$$I = -\int_{\pi/2}^0 \frac{\cos^n\left(\frac{\pi}{2} - t\right)}{\sin^n\left(\frac{\pi}{2} - t\right) + \cos^n\left(\frac{\pi}{2} - t\right)} dt = \int_0^{\pi/2} \frac{\sin^n t}{\sin^n t + \cos^n t} dt. \qquad (13.5)$$

By summing (13.4) and (13.5), we get

$$2I = \int_0^{\pi/2} \frac{\sin^n t + \cos^n t}{\sin^n t + \cos^n t} dt = \int_0^{\pi/2} dt = \frac{\pi}{2} \implies I = \frac{\pi}{4}.$$

In a similar way, we also find

$$I = \int_0^{\pi/2} \frac{\sin^n x}{\sin^n x + \cos^n x} dx = \frac{\pi}{4}, \qquad \forall n \geqslant 0.$$

Now, for the given integral $I_{15}$, we have

$$I_{15} = \int_0^{\pi/2} \frac{\cos^{\sqrt{2020}} x}{\sin^{\sqrt{2020}} x + \cos^{\sqrt{2020}} x} dx = \frac{\pi}{4}.$$

$$I_{16} = \int x(1-x)^{2020}\,dx$$

**Solution 1.**

For the given integral we can write the following

$$I_{16} = -\int -x(1-x)^{2020}\,dx = -\int (1-x-1)(1-x)^{2020}\,dx$$

$$= -\int \left((1-x)^{2021} - (1-x)^{2020}\right)dx = \int \left((1-x)^{2020} - (1-x)^{2021}\right)dx$$

$$= \frac{(1-x)^{2022}}{2022} - \frac{(1-x)^{2021}}{2021} + C,$$

where $C \in \mathbb{R}$.

**Solution 2.**

Let us use the substitution $1-x = t$, so that $dx = -dt$. This gives the following

$$I_{16} = -\int (1-t)t^{2020}\,dt = \int \left(t^{2021} - t^{2020}\right)dt = \frac{t^{2022}}{2022} - \frac{t^{2021}}{2021} + C$$

$$= \frac{(1-x)^{2022}}{2022} - \frac{(1-x)^{2021}}{2021} + C,$$

where $C \in \mathbb{R}$.

**Solution 3.**

By using the integration by parts, let us assume

$$u = x \implies du = dx, \quad dv = (1-x)^{2020}\,dx \implies v = -\frac{(1-x)^{2021}}{2021}.$$

Thus, we have

$$I_{16} = -\frac{x(1-x)^{2020}}{2021} + \frac{1}{2021}\int (1-x)^{2021}\,dx$$

$$= -\frac{x(1-x)^{2021}}{2021} - \frac{(1-x)^{2022}}{4086462} + C,$$

where $C \in \mathbb{R}$.

$$I_{17} = \int \frac{\sec^4 x \tan x}{\sec^4 x + 4}\,dx$$

**Solution 1.**

Let us use the substitution $\sec^4 x + 4 = t$, so that

$$4\sec^3 x \left(\frac{1}{\cos x}\right)' dx = dt \implies 4\sec^3 x \frac{\sin x}{\cos^2 x} dx = dt$$

$$\implies 4\sec^3 x \left(\frac{1}{\cos x}\right) \tan x\, dx = dt$$

$$\implies \sec^4 x \tan x\, dx = dt.$$

This gives the following

$$I_{16} = \frac{1}{4}\int \frac{dt}{t} = \frac{1}{4}\ln|t| + C = \frac{1}{4}\ln\left(4 + \sec^4 x\right) + C,$$

where $C \in \mathbb{R}$.

**Solution 2.**

For the given integral we can write the following

$$I_{17} = \int \frac{\frac{1}{\cos^4 x}\frac{\sin x}{\cos x}}{\frac{1}{\cos^4 x} + 4} dx = \int \frac{\sin x}{\cos^5 x \left(\frac{1 + 4\cos^4 x}{\cos^4 x}\right)} dx$$

$$= \int \frac{\sin x}{\cos x\,(1 + 4\cos^4 x)} dx.$$

For the last integral, let us use the substitution $\cos x = t$, so that $\sin x\, dx = -dt$. This gives the following

$$I_{17} = -\int \frac{dt}{t(1 + 4t^4)} = -\int \frac{dt}{t^5\left(4 + \frac{1}{t^4}\right)} = -\int \frac{1}{t^2\left(4 + \frac{1}{t^4}\right)}\frac{dt}{t^3}.$$

Now, let us use the substitution $1/t^2 = u$, so that $dt/t^3 = -du/2$. This gives the following

$$I_{17} = \frac{1}{2}\int \frac{u}{4 + u^2} du = \frac{1}{4}\int \frac{2u}{4 + u^2} du = \frac{1}{4}\ln(4 + u^2) + C$$

$$= \frac{1}{4}\ln\left(4 + \frac{1}{t^4}\right) + C = \frac{1}{4}\ln\left(4 + \frac{1}{\cos^4 x}\right) + C = \frac{1}{4}\ln\left(4 + \sec^4 x\right) + C,$$

where $C \in \mathbb{R}$.

$$\boxed{I_{18} = \int x^{2x}\,(2 + 2\ln x)\, dx}$$

**Solution.**

Let us use the substitution $x^{2x} = t$, so that $2x \ln x = \ln t$, and $(2 + 2 \ln x)dx = dt/t$. This gives the following

$$I_{18} = \int t \frac{dt}{t} = \int dt = t + C = x^{2x} + C,$$

where $C \in \mathbb{R}$.

$$\boxed{I_{19} = \int_0^1 \sqrt{1 - x^2} dx}$$

**Solution 1.**

Let us use the substitution $x = \sin t$, so that $dx = \cos t\, dt$. If $x = 1$, then $t = 0$, and if $x = 1$, then $t = \pi/2$. With these new values, we get the following

$$I_{19} = \int_0^{\pi/2} \sqrt{1 - \sin^2 t} \cos t\, dt = \int_0^{\pi/2} |\cos t| \cos t\, dt = \int_0^{\pi/2} \cos^2 t\, dt$$

$$= \frac{1}{2} \int_0^{\pi/2} (1 + \cos(2t))dt = \frac{1}{2} \left[ t + \frac{1}{2} \sin(2t) \right]_0^{\pi/2} = \frac{1}{2} \left( \frac{\pi}{2} - 0 \right) = \frac{\pi}{4}.$$

**Solution 2.**

From the geometrical viewpoint, the given integral represents the area of the first quarter of the unit circle, its equation is $x^2 + y^2 = 1$. Therefore, $I_{19} = \pi/4$.

$$\boxed{I_{20} = \int_0^\infty x^5 e^{-x^4} dx}$$

**Solution 1.**

For the given integral we can write the following

$$I_{20} = \int_0^\infty x(x^2)^2 e^{-(x^2)^2} dx.$$

Let us use the substitution $x^2 = t$, so that $2x\, dx = dt$. If $x = 0$, then $t = 0$, and if $x \to \infty$, then $t \to \infty$. This gives the following

$$I_{20} = \frac{1}{2} \int_0^\infty t^2 e^{-t^2} dt.$$

Note that the function $f(w) := w^2 e^{-w^2}$ is an even function on $\mathbb{R} = ]-\infty, \infty[$. Thus, we get

$$\int_{-\infty}^\infty w^2 e^{-w^2} dw = 2 \int_0^\infty w^2 e^{-w^2} dw.$$

Therefore, we can write

$$I_{20} = \frac{1}{4} \int_{-\infty}^{\infty} t^2 e^{-t^2} dt = \frac{1}{4} \int_{-\infty}^{\infty} t \left( t e^{-t^2} \right) dt.$$

By using the integration by parts, let us assume

$$u = t \Longrightarrow du = dt, \quad dv = t e^{-t^2} dt \Longrightarrow v = -\frac{1}{2} e^{-t^2}.$$

Thus, we have

$$I_{20} = \frac{1}{4} \left( \lim_{a \to \infty} \left[ -\frac{1}{2} t e^{-t^2} \right]_{-a}^{a} + \frac{1}{2} \int_{-\infty}^{\infty} e^{-t^2} dt \right)$$

$$= \frac{1}{4} \left( -\frac{1}{2} \lim_{a \to \infty} \left( 2a e^{-a^2} \right) + \frac{\sqrt{\pi}}{2} \right) = \frac{\sqrt{\pi}}{8}.$$

Where we used the Gaussian integral (1.14).

**Solution 2.**
Let us calculate the given integral $I_{20}$, by using the gamma function, where

$$\Gamma(a) = \int_{0}^{\infty} x^{a-1} e^{-x} dx, \quad \forall a > 0.$$

Let us use the substitution $x^4 = t$, so that $x^2 = \sqrt{t}$, and $4x^3 dx = dt$. If $x = 0$, then $t = 0$, and if $x \to \infty$, then $t \to \infty$. This gives the following

$$I_{20} = \frac{1}{4} \int_{0}^{\infty} \sqrt{t} e^{-t} dt = \frac{1}{4} \int_{0}^{\infty} t^{\frac{3}{2}-1} e^{-t} dt = \frac{1}{4} \Gamma \left( \frac{3}{2} \right) = \frac{1}{4} \frac{1}{2} \Gamma \left( \frac{1}{2} \right) = \frac{\sqrt{\pi}}{8}.$$

Remember that $\Gamma(a+1) = a\Gamma(a)$ and $\Gamma(1/2) = \sqrt{\pi}$ (see Subsec. 1.8.1).

# Chapter 14

# The Solutions to the 2022 MIT Integration Bee, Qualifying Test

$$I_1 = \int \frac{1 + \cos x}{x + \sin x} dx$$

**Solution.**
This integral is simple, where $(x+\sin x)' = 1+\cos x$. Thus $I_1 = \ln |x + \sin x| + C$, where $C \in \mathbb{R}$.

$$I_2 = \int_1^{\sqrt{3}} \frac{\arctan x + \operatorname{arccot} x}{x} dx$$

**Solution.**
We have

$$\arctan x + \operatorname{arccot} x = \frac{\pi}{2}, \quad \forall x \in \mathbb{R}. \tag{14.1}$$

To prove this identity, we can write the following.
**Proof 1.**
Let $f(x) := \arctan x + \operatorname{arccot} x$. Then $f'(x) = \frac{1}{1+x^2} - \frac{1}{1+x^2} = 0$. Thus, the function $f$ is a constant, i.e. $f(x) = c \in \mathbb{R}$. Let us substitute $x = 1$, then $c = f(1) = \arctan(1) + \operatorname{arccot}(1) = \frac{\pi}{4} + \frac{\pi}{4} = \frac{\pi}{2}$. Therefore we have $f(x) = \frac{\pi}{2}$, $\forall x \in \mathbb{R}$.
**Proof 2.**
Let $\theta := \arctan x$, then $x = \tan \theta$, and

$$x = \cot\left(\frac{\pi}{2} - \theta\right) \implies \operatorname{arccot} x = \frac{\pi}{2} - \theta \implies \operatorname{arccot} x + \theta = \frac{\pi}{2}$$

$$\implies \arctan x + \operatorname{arccot} x = \frac{\pi}{2}.$$

Therefore, for the given integral $I_2$, we have

$$I_2 = \frac{\pi}{2} \int_1^{\sqrt{3}} \frac{dx}{x} = \frac{\pi}{2} \left[ \ln x \right]_1^{\sqrt{3}} = \frac{\pi}{2} \ln(\sqrt{3}) = \frac{\pi \ln 3}{4}.$$

$$I_3 = \int_0^{2022} \left( x^2 - \lfloor x \rfloor \lceil x \rceil \right) dx$$

**Solution.**

Let us find a closed formula for the integral

$$J(n) = \int_0^n \left( x^2 - \lfloor x \rfloor \lceil x \rceil \right) dx \quad n \in \mathbb{N} = \{1, 2, \ldots\},$$

as a function of $n$. Then the given integral $I_3$ will be $J(2022)$.
We have

$$\lfloor x \rfloor = 0, \quad \lceil x \rceil = 1, \quad \forall\, 0 \leqslant x < 1,$$
$$\lfloor x \rfloor = 1, \quad \lceil x \rceil = 2, \quad \forall\, 1 \leqslant x < 2,$$
$$\lfloor x \rfloor = 2, \quad \lceil x \rceil = 3, \quad \forall\, 2 \leqslant x < 3, \quad \ldots$$
$$\lfloor x \rfloor = n - 1, \quad \lceil x \rceil = n, \quad \forall\, n - 1 \leqslant x < n.$$

Thus, we have

$$J(n) = \int_0^n \left( x^2 - \lfloor x \rfloor \lceil x \rceil \right) dx = \int_0^n x^2 dx - \int_0^n \lfloor x \rfloor \lceil x \rceil dx$$

$$= \int_0^n x^2 dx - \left( \int_0^1 (0)(1) dx + \int_1^2 (1)(2) dx + \int_2^3 (2)(3) dx + \ldots + \right.$$
$$\left. + \int_{n-1}^n (n-1) n dx \right)$$

$$= \left[ \frac{x^3}{3} \right]_0^n - \sum_{k=1}^n \left( \int_{k-1}^k k(k-1) dx \right)$$

$$= \frac{n^3}{3} - \sum_{k=1}^n k(k-1)(k-k+1) = \frac{n^3}{3} - \sum_{k=0}^n k^2 + \sum_{k=1}^n k$$

$$= \frac{n^3}{3} - \left( 1^2 + 2^2 + 3^2 + \ldots + n^2 \right) + \left( 1 + 2 + \ldots + n \right)$$

$$= \frac{n^3}{3} - \frac{n(n+1)(2n+1)}{6} + \frac{n(n+1)}{2} = \frac{n}{3}.$$

Therefore, for the given integral $I_3$, we get

$$I_3 = J(2022) = \frac{2022}{3} = 674.$$

$$I_4 = \int \frac{\sinh x}{\cosh x - \sinh x} dx$$

**Solution.**

For the given integral we can write the following

$$
I_4 = \int \frac{\frac{e^x - e^{-x}}{2}}{\frac{e^x + e^{-x}}{2} - \frac{e^x - e^{-x}}{2}} dx = \int \frac{e^x - e^{-x}}{e^x + e^{-x} - e^x + e^{-x}} dx = \int \frac{e^x - e^{-x}}{2e^{-x}} dx
$$

$$
= \frac{1}{2} \int \left( e^x - e^{-x} \right) e^x dx = \frac{1}{2} \int \left( e^{2x} - 1 \right) dx = \frac{1}{2} \left( \frac{e^{2x}}{2} - x \right) + C
$$

$$
= -\frac{x}{2} + \frac{e^{2x}}{4} + C,
$$

where $C \in \mathbb{R}$.

$$I_5 = \int \frac{x}{\sqrt{x-1} + \sqrt{x+1}} dx$$

**Solution.**

For the given integral we can write the following

$$
I_5 = \int \frac{x \left( \sqrt{x-1} - \sqrt{x+1} \right)}{(x-1) - (x+1)} dx = -\frac{1}{2} \int \left( x\sqrt{x-1} - x\sqrt{x+1} \right) dx
$$

$$
= -\frac{1}{2} \int \left( (x - 1 + 1)(x-1)^{1/2} - (x + 1 - 1)(x+1)^{1/2} \right) dx
$$

$$
= -\frac{1}{2} \int \left( (x-1)^{3/2} + (x-1)^{1/2} - (x+1)^{3/2} + (x+1)^{1/2} \right) dx
$$

$$
= -\frac{1}{2} \left( \frac{2}{5}(x-1)^{5/2} + \frac{2}{3}(x-1)^{3/2} - \frac{2}{5}(x+1)^{5/2} + \frac{2}{3}(x+1)^{3/2} \right) + C
$$

$$
= \frac{(x+1)^{5/2}}{5} - \frac{(x+1)^{3/2}}{3} - \frac{(x-1)^{5/2}}{5} - \frac{(x-1)^{3/2}}{3} + C,
$$

where $C \in \mathbb{R}$.

$$I_6 = \int_0^\pi \cos(x + \cos x) dx$$

**Solution.**

Let us use the substitution $\pi - x = t$, so that $dx = -dt$. If $x = 0$, then $t = \pi$, and if $x = \pi$, then $t = 0$. With these new values, we get the following

$$I_6 = -\int_\pi^0 \cos\left(\pi - t + \cos(\pi - t)\right) dt = \int_0^\pi \cos\left(\pi - (t + \cos t)\right) dt$$

$$= -\int_0^\pi \cos(t + \cos t)dt = -I_6.$$

Thus, $2I_6 = 0$, i.e. $I_6 = 0$.

**Remark.** The proposed substitution in the solution of $I_6$ is equivalent to the use of King property (1.13).

$$\boxed{I_7 = \int x^3 \sin(x^2)dx}$$

**Solution.**

For the given integral we can write the following

$$I_7 = \int xx^2 \sin(x^2)dx.$$

Let us use the substitution $x^2 = t$, so that $2xdx = dt$. This gives the following

$$I_7 = \frac{1}{2}\int t \sin t\, dt.$$

Now, by using the integration by parts (assume $u = t, dv = \sin t dt$), we get

$$I_7 = \frac{1}{2}\left(-t\cos t + \int \cos t dt\right) = -\frac{t\cos t}{2} + \frac{\sin t}{2} + C$$

$$= \frac{\sin(x^2) - x^2 \cos(x^2)}{2} + C,$$

where $C \in \mathbb{R}$.

$$\boxed{I_8 = \int \frac{x}{1 - x^4}dx}$$

**Solution.**

Let us use the substitution $x^2 = t$, so that $2xdx = dt$. This gives the following

$$I_8 = \frac{1}{2}\int \frac{dt}{1 - t^2} = \frac{1}{2}\frac{1}{2}\ln\left|\frac{1 + t}{1 - t}\right| + C = \frac{1}{4}\ln\left|\frac{1 + x^2}{1 - x^2}\right| + C,$$

where $C \in \mathbb{R}$.

$$I_9 = \int \frac{dx}{\cosh^2 x}$$

### Solution 1.

This integral is simple, where $I_9 = \tanh x + C$, and $C \in \mathbb{R}$.

### Solution 2.

For the given integral we can write the following

$$I_9 = \int \frac{dx}{\left(\frac{e^x + e^{-x}}{2}\right)^2} = 4 \int \frac{dx}{(e^x + e^{-x})^2} = 4 \int \frac{dx}{e^{-2x}(e^{2x} + 1)^2}$$

$$= 2 \int 2e^{2x}(e^{2x} + 1)^{-2} dx = -\frac{2}{e^{2x} + 1} + C,$$

where $C \in \mathbb{R}$.

$$I_{10} = \int_0^1 \left(e^{e^x} - e^{e^x - x}\right) dx$$

### Solution 1.

We note that

$$\left(\frac{e^{e^x}}{e^x}\right)' = \frac{e^{e^x}(e^x)^2 - e^x e^{e^x}}{(e^x)^2} = e^{e^x} - e^{e^x} e^{-x} = e^{e^x} - e^{e^x - x}.$$

Thus, for the given integral we get

$$I_{10} = \int_0^1 \left(\frac{e^{e^x}}{e^x}\right)' dx = \left[\frac{e^{e^x}}{e^x}\right]_0^1 = \left(\frac{e^e}{e} - e\right) = e^{e-1} - e.$$

### Solution 2.

For the given integral we can write the following

$$I_{10} = \int_0^1 \left(e^{e^x} - e^{e^x} e^{-x}\right) dx.$$

Let us use the substitution $e^x = t$, so that $e^x dx = dt$, thus $dx = dt/t$. If $x = 0$, then $t = 1$, and if $x = 1$, then $t = e$. With these new values, we get the following

$$I_{10} = \int_1^e \left(e^t - \frac{e^t}{t}\right) \frac{dt}{t} = \int_1^e \left(\frac{e^t}{t} - \frac{e^t}{t^2}\right) dt = \underbrace{\int_1^e \frac{e^t}{t} dt}_{:=J} - \int_1^e \frac{e^t}{t^2} dt.$$

For the integral $J$, by using the integration by parts, let us assume

$$u = \frac{1}{t} \implies du = -\frac{dt}{t^2}, \quad dv = e^t dt \implies v = e^t.$$

Thus, we have

$$I_{10} = \left[\frac{e^t}{t}\right]_1^e + \int_1^e \frac{e^t}{t^2} dt - \int_1^e \frac{e^t}{t^2} dt = \frac{e^e}{e} - e = e^{e-1} - e.$$

$$I_{11} = \lim_{n \to \infty} \int_0^3 \sin\left(\frac{\pi}{3} \sin\left(\frac{\pi}{3} \sin\left(\ldots \sin\left(\frac{\pi}{3}x\right)\right)\right)\right) dx$$
$$\underbrace{\qquad\qquad\qquad\qquad\qquad\qquad\qquad\qquad\qquad}_{n \text{ sin's}}$$

**Solution.**

First of all, let us understand the construction of the integrand in $I_{11}$ (see Fig. 14.1). Let $x \neq 0$, $g(x) := \sin\left(\frac{\pi}{3}x\right)$, and

$$g_1(x) := g(x) = \sin\left(\frac{\pi}{3}x\right),$$
$$g_2(x) := g \circ g(x) = g(g(x)) = \sin\left(\frac{\pi}{3}\sin\left(\frac{\pi}{3}x\right)\right),$$
$$g_3(x) := g \circ g \circ g(x) = \sin\left(\frac{\pi}{3}\sin\left(\frac{\pi}{3}\sin\left(\frac{\pi}{3}x\right)\right)\right),$$
$$g_4(x) := g \circ g \circ g \circ g(x) = \sin\left(\frac{\pi}{3}\sin\left(\frac{\pi}{3}\sin\left(\frac{\pi}{3}\sin\left(\frac{\pi}{3}x\right)\right)\right)\right), \quad \ldots$$
$$g_n(x) := \underbrace{g \circ g \ldots \circ g}_{n \text{ times}}(x) = \underbrace{\sin\left(\frac{\pi}{3}\sin\left(\frac{\pi}{3}\sin\left(\ldots \sin\left(\frac{\pi}{3}x\right)\right)\right)\right)}_{n \text{ sin's}}.$$

Therefore, we have

$$I_{11} = \lim_{n \to \infty} \int_0^3 f_n(x) dx = \int_0^3 \lim_{n \to \infty} \underbrace{\sin\left(\frac{\pi}{3}\sin\left(\frac{\pi}{3}\sin\left(\ldots \sin\left(\frac{\pi}{3}x\right)\right)\right)\right)}_{n \text{ sin's}} dx$$
$$= \int_0^3 \underbrace{\sin\left(\frac{\pi}{3}\sin\left(\frac{\pi}{3}\sin\left(\ldots \sin\left(\frac{\pi}{3}x\right)\ldots\right)\right)\right)}_{\infty \text{ sin's}} dx.$$

Let us denote

$$\underbrace{\sin\left(\frac{\pi}{3}\sin\left(\ldots \sin\left(\frac{\pi}{3}x\right)\ldots\right)\right)}_{\infty \text{ sin's}} := t.$$

Then, we have

$$f(x) := \sin\left(\frac{\pi}{3}\underbrace{\sin\left(\frac{\pi}{3}\sin\left(\ldots\sin\left(\frac{\pi}{3}x\right)\ldots\right)\right)}_{t}\right) = t,$$

i.e.

$$\sin\left(\pi t/3\right) = t. \tag{14.2}$$

It is obvious that $t = 1/2 \in ]0,3[$ is a solution of (14.2). Thus $f(x) = 1/2$. Therefore, we have

$$I_{11} = \int_0^3 \frac{1}{2}dx = \frac{3}{2}.$$

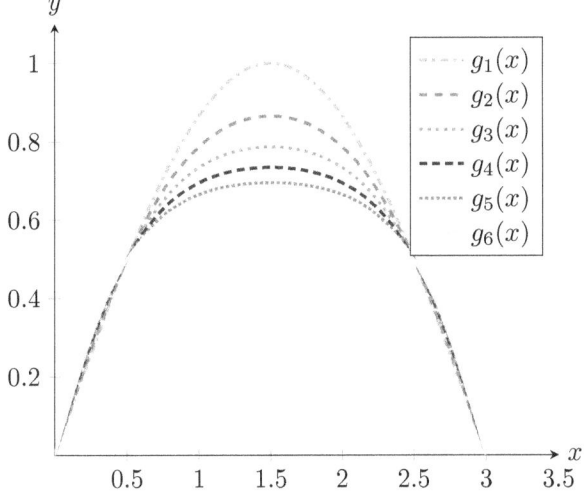

Figure 14.1: The graphs of $g_1, g_2, g_3, g_4, g_5$ and $g_6$ in $[0,3]$.

$$I_{12} = \int_0^1 \sqrt{1 - \sqrt{x}}dx$$

**Solution 1.**

Let us use the substitution $x = t^2$, so that $dx = 2tdt$. If $x = 0$, then $t = 0$, and if $x = 1$, then $t = 1$. This gives the following

$$I_{12} = \int_0^1 \sqrt{1 - |t|}(2t)dt = 2\int_0^1 \sqrt{1 - t^2}\, t\, dt$$

$$= -2 \int_0^1 (-t + 1 - 1)(1 - t)^{1/2} dt$$

$$= -2 \int_0^1 \left( (1 - t)^{3/2} - (1 - t)1/2 \right) dt$$

$$= -2 \left[ -\frac{2}{5}(1 - t)^{5/2} + \frac{2}{3}(1 - t)^{3/2} \right]_0^1$$

$$= -2 \left( \frac{2}{5} - \frac{2}{3} \right) = \frac{8}{15}.$$

**Solution 2.**

Let us use the substitution $1 - \sqrt{x} = t$, so that $x = (1 - t)^2$ and $dx = -2(1 - t)dt$. If $x = 0$, then $t = 1$, and if $x = 1$, then $t = 0$. This gives the following

$$I_{12} = -2 \int_1^0 \sqrt{t}(1 - t)dt = 2 \int_0^1 \left( t^{1/2} - t^{3/2} \right) dt = 2 \left[ \frac{2}{3}t^{3/2} - \frac{2}{5}t^{5/2} \right]_0^1$$

$$= 2 \left( \frac{2}{3} - \frac{2}{5} \right) = \frac{8}{15}.$$

$$\boxed{I_{13} = \int \frac{x^3}{1 + x + \frac{x^2}{2} + \frac{x^3}{6}} dx}$$

**Solution.**

For the given integral we can write the following

$$I_{13} = \int \frac{6x^3}{x^3 + 3x^2 + 6x + 6} dx = 6 \int \left( 1 - \frac{3x^2 + 6x + 6}{x^3 + 3x^2 + 6x + 6} \right) dx$$

$$= 3 \left( x - \ln \left| x^3 + 3x^2 + 6x + 6 \right| \right) + C,$$

where $C \in \mathbb{R}$.

$$\boxed{I_{14} = \int \left( \sin(x + \sin x) - \sin(x - \sin x) \right) dx}$$

**Solution.**

For the given integral we can write the following

$$I_{14} = \int \Big( \sin x \cos(\sin x) + \cos x \sin(\sin x) - \sin x \cos(\sin x)$$

$$+ \cos x \sin(\sin x) \Big) dx$$

$$= 2 \int \cos x \sin(\sin x) dx.$$

Let us use the substitution $\sin x = t$, so that $\cos x\, dx = dt$. This gives the following

$$I_{14} = 2 \int \sin t\, dt = -2\cos t + C = -2\cos(\sin x) + C,$$

where $C \in \mathbb{R}$.

$$\boxed{I_{15} = \int \left(\tan^4 x \sec^3 x + \tan^2 x \sec^5 x\right) dx}$$

**Solution 1.**

For the given integral we can write the following

$$
\begin{aligned}
I_{15} &= \int \left(\sec^3 x (\tan^2 x)^2 + \sec^5 x (\sec^2 x - 1)\right) dx \\
&= \int \left(\sec^3 x (\sec^2 x - 1)^2 + \sec^5 x (\sec^2 x - 1)\right) dx \\
&= \int \left(\sec^3 x (\sec^4 x - 2\sec^2 x + 1) + \sec^7 x - \sec^5 x\right) dx \\
&= \int \left(\sec^7 x - 2\sec^5 x + \sec^3 x + \sec^7 x - \sec^5 x\right) dx \\
&= \int \left(2\sec^7 x - 3\sec^5 x + \sec^3 x\right) dx \\
&= 2 \int \sec^7 x\, dx - 3 \int \sec^5 x\, dx + \int \sec^3 x\, dx.
\end{aligned}
$$

By applying the following reduction formula

$$\int \cos^n x\, dx := J_n = \frac{n+2}{n+1} J_{n+2} - \frac{\sin x \cos^{n+1} x}{n+1}, \quad n = -2, -3, -4, \ldots$$

we find

$$
\begin{aligned}
I_{15} &= 2J_{-7} - 3J_{-5} + J_{-3} = 2\left(\frac{5}{6}J_{-5} + \frac{\sin x}{6\cos^6 x}\right) - 3J_{-5} + J_{-3} \\
&= \frac{5}{3}J_{-5} + \frac{\sin x}{3\cos^6 x} - 3J_{-5} + J_{-3} = -\frac{4}{3}J_{-5} + \frac{\sin x}{3\cos^6 x} + J_{-3} \\
&= -\frac{4}{3}\left(\frac{3}{4}J_{-3} + \frac{\sin x}{4\cos^4 x}\right) + \frac{\sin x}{3\cos^6 x} + J_{-3} \\
&= -J_{-3} - \frac{\sin x}{3\cos^4 x} + \frac{\sin x}{3\cos^6 x} + J_{-3}
\end{aligned}
$$

$$= \frac{\sin x - \sin x(1 - \sin^2 x)}{3\cos^6 x} + C = \frac{\sin x - \sin x \cos^2 x}{3\cos^6 x} + C$$

$$= \frac{\sin^3 x}{3\cos^6 x} + C = \frac{1}{3}\frac{\sin^3 x}{\cos^3 x}\frac{1}{\cos^3 x} + C = \frac{1}{3}\tan^3 x \sec^3 x + C,$$

where $C \in \mathbb{R}$.

## Solution 2.

Because there is a similarity between the terms of the integrand, we will use the following property

$$f(x)g(x) = \int (f(x)g(x))' \, dx = \int (f'(x)g(x) + f(x)g'(x)) \, dx.$$

Therefore, let us find $\alpha, \beta, \gamma$ in $f(x)g(x) = \alpha \tan^\beta x \sec^\gamma x$, such that

$$\left(\alpha \tan^\beta x \sec^\gamma x\right)' = \tan^4 x \sec^3 x + \tan^2 x \sec^5 x,$$

i.e.

$$\alpha\beta \tan^{\beta-1} x \sec^\gamma x \underbrace{\sec^2 x}_{(\tan x)'} + \alpha\gamma \tan^\beta x \sec^{\gamma-1} x \underbrace{\tan x \sec x}_{(\sec x)'}$$

$$= \tan^4 x \sec^3 x + \tan^2 x \sec^5 x.$$

i.e.

$$\alpha\beta \tan^{\beta-1} x \sec^{\gamma+2} x + \alpha\gamma \tan^{\beta+1} x \sec^\gamma x = \tan^4 x \sec^3 x + \tan^2 x \sec^5 x. \tag{14.3}$$

By corresponding between tow sides of (14.3), we find

$$\alpha\beta \tan^{\beta-1} x \sec^{\gamma+2} x = \tan^2 x \sec^5 x, \tag{14.4}$$

and

$$\alpha\gamma \tan^{\beta+1} x \sec^\gamma x = \tan^4 x \sec^3 x. \tag{14.5}$$

From (14.4), we find

$$\beta - 1 = 2, \quad \gamma + 2 = 5, \quad \alpha\beta = 1 \Longrightarrow \alpha = 1/3, \quad \beta = 3, \quad \gamma = 3.$$

Note that $\alpha = 1/3, \beta = 3, \gamma = 3$ satisfy (14.5). As a result, we get the following

$$\left(\frac{1}{3}\tan^3 x \sec^3 x\right)' = \tan^4 x \sec^3 x + \tan^2 x \sec^5 x.$$

Therefore, for the given integral $I_{15}$, we have

$$I_{15} = \int \left(\frac{1}{3}\tan^3 x \sec^3 x\right)' \, dx = \frac{1}{3}\tan^3 x \sec^3 x + C,$$

where $C \in \mathbb{R}$.

$$I_{16} = \int (1 + \ln x) \ln (\ln x)\, dx$$

**Solution.**
For the given integral we can write the following

$$I_{16} = \int \ln (\ln x)\, dx + \underbrace{\int \ln x \ln (\ln x)\, dx}_{:=J}.$$

For the integral $J$, by using the integration by parts, let us assume

$$u = \ln x \ln (\ln x) \implies du = \left(\frac{\ln(\ln x)}{x} + \frac{1}{x}\right) dx, \quad dv = dx \implies v = x.$$

Thus, we have

$$I_{16} = \int \ln (\ln x)\, dx + x \ln x \ln (\ln x) - \int x \left(\frac{\ln(\ln x)}{x} + \frac{1}{x}\right) dx$$
$$= \int \ln (\ln x)\, dx + x \ln x \ln (\ln x) - \int \ln (\ln x)\, dx - \int dx$$
$$= x \ln x \ln (\ln x) - x + C,$$

where $C \in \mathbb{R}$.

$$I_{17} = \int \Big(\frac{1}{1 + \sin x} + \frac{1}{1 + \cos x} + \frac{1}{1 + \tan x} + \frac{1}{1 + \cot x}$$
$$+ \frac{1}{1 + \sec x} + \frac{1}{1 + \csc x}\Big) dx.$$

**Solution.**
For the given integral we can write the following

$$I_{17} = \int \Big(\frac{1}{1 + \sin x} + \frac{1}{1 + \cos x} + \frac{\cos x}{\sin x + \cos x} + \frac{\sin x}{\sin x + \cos x}$$
$$+ \frac{\cos x}{1 + \cos x} + \frac{\sin x}{1 + \sin x}\Big) dx$$
$$= \int \Big(\frac{1 + \sin x}{1 + \sin x} + \frac{1 + \cos x}{1 + \cos x} + \frac{\sin x + \cos x}{\sin x + \cos x}\Big) dx$$
$$= \int 3 dx = 3x + C,$$

where $C \in \mathbb{R}$.

$$I_{18} = \int \frac{dx}{\sqrt{x - x^2}}$$

### Solution 1.

For the given integral we can write the following

$$I_{18} = \int \frac{dx}{\sqrt{x(1-x)}} = \int \frac{dx}{\sqrt{x}\sqrt{1-x}}.$$

For the last integral, let us use the substitution $x = t^2$, so that $dx = 2t\,dt$. This gives the following

$$I_{18} = 2 \int \frac{t}{t\sqrt{1-t^2}}\,dt = 2 \int \frac{dt}{\sqrt{1-t^2}} = 2\arcsin t + C = 2\arcsin(\sqrt{x}) + C,$$

where $C \in \mathbb{R}$.

### Solution 2.

For the given integral we can write the following

$$I_{18} = \int \frac{dx}{\sqrt{-\left(x^2 - x + \frac{1}{4} - \frac{1}{4}\right)}} = \int \frac{dx}{\sqrt{\frac{1}{4} - \left(x - \frac{1}{2}\right)^2}}$$

$$= \arcsin\left(\frac{x - 1/2}{1/2}\right) + C = \arcsin(2x - 1) + C,$$

where $C \in \mathbb{R}$.

$$I_{19} = \int_0^{1/2} \left( \sum_{n=0}^{\infty} \binom{n+3}{n} x^n \right) dx$$

### Solution.

For the given integral we can write the following

$$I_{19} = \int_0^{1/2} \left( \sum_{n=0}^{\infty} \frac{(n+3)!}{(n!)(n+3-n)!} x^n \right) dx$$

$$= \int_0^{1/2} \left( \sum_{n=0}^{\infty} \frac{(n+3)(n+2)(n+1)}{3!} x^n \right) dx$$

$$= \frac{1}{6} \int_0^{1/2} \left( \sum_{n=0}^{\infty} (n+3)(n+2)(n+1) x^n \right) dx.$$

Let us set $a_n := (n+3)(n+2)(n+1)x^n$, then

$$\lim_{n\to\infty}\left|\frac{a_{n+1}}{a_n}\right| = \lim_{n\to\infty}\frac{(n+4)(n+3)(n+2)}{(n+3)(n+2)(n+1)}\left|\frac{x^{n+1}}{x^n}\right| = |x|.$$

Therefore, by the D'Alembert's Ratio Test, the power series (which is the integrand in $I_{19}$) $\sum_{n=0}^{\infty}(n+3)(n+2)(n+1)x^n$ is absolutely convergence for any $x \in ]-1,1[$. We know that a power series representation of any function $f(x)$ can be integrated term-by-term from $a$ to $b$ to obtain a series representation of the definite integral $\int_a^b f(x)dx$, provided that $]a,b[$ lies within the interval of the convergence of the power series that represents $f(x)$. Thus, because $]0,1/2[ \subset ]-1,1[$, we find

$$I_{19} = \frac{1}{6}\sum_{n=0}^{\infty}\left(\int_0^{1/2}(n+3)(n+2)(n+1)x^n dx\right)$$

$$= \frac{1}{6}\sum_{n=0}^{\infty}\left([(n+3)(n+1)x^{n+1}]_{x=0}^{x=1/2}\right)$$

$$= \frac{1}{6}\sum_{n=0}^{\infty}\frac{(n+3)(n+2)}{2^{n+1}} = \frac{1}{12}\sum_{n=0}^{\infty}\frac{n^2+5n+6}{2^n}$$

$$= \frac{1}{12}\left(\underbrace{\sum_{n=1}^{\infty}\frac{n^2}{2^n}}_{:=S_1} + 5\underbrace{\sum_{n=1}^{\infty}\frac{n}{2^n}}_{:=S_2} + 6\underbrace{\sum_{n=0}^{\infty}\frac{1}{2^n}}_{:=S_3}\right).$$

Simply we find that $S_3 = \frac{1}{1-\frac{1}{2}} = 2$ (the sum of geometric series).

For $S_2$, let

$$h(x) := \sum_{n=1}^{\infty}x^n = \frac{x}{1-x}, \quad -1 < x < 1.$$

Thus,

$$h'(x) = \frac{1}{(1-x)^2} = \sum_{n=1}^{\infty}nx^{n-1}, \quad -1 < x < 1. \tag{14.6}$$

When $x = 1/2$, we find

$$\frac{1}{4} = \sum_{n=1}^{\infty}\frac{n}{2^{n-1}} = 2\sum_{n=1}^{\infty}\frac{n}{2^n} \implies S_2 = \sum_{n=1}^{\infty}\frac{n}{2^n} = 2.$$

For $S_1$, from (14.6) we get

$$(xh'(x))' = \left(\frac{x}{(1-x)^2}\right)' = \left(\sum_{n=1}^{\infty}nx^n\right)' = \sum_{n=1}^{\infty}n^2 x^{n-1}.$$

Thus,

$$\frac{1+x}{(1-x)^3} = \sum_{n=1}^{\infty} \frac{n^2 x^n}{x} \implies \frac{x(1+x)}{(1-x)^3} = \sum_{n=1}^{\infty} n^2 x^n.$$

When $x = 1/2$, we get

$$S_1 = \sum_{n=1}^{\infty} \frac{n^2}{2^n} = \frac{3/2}{1/8} = 6.$$

Therefore, for the given integral $I_{19}$, we find

$$I_{19} = \frac{1}{12} \left(6 + 5(2) + 6(2)\right) = \frac{7}{3}.$$

$$\boxed{I_{20} = \int \frac{dx}{1 + \cos^2 x}}$$

**Solution.**
For the given integral we can write the following

$$I_{20} = \int \frac{1/\cos^2 x}{1 + \frac{1}{\cos^2 x}} dx = \int \frac{1/\cos^2 x}{2 + \left(\frac{1}{\cos^2 x} - 1\right)} dx = \int \frac{1/\cos^2 x}{2 + \tan^2 x} dx.$$

Let us use the substitution $\tan x = t$, so that $dx/\cos^2 x = dt$. This gives the following

$$I_{20} = \int \frac{dt}{2 + t^2} = \frac{1}{\sqrt{2}} \arctan\left(\frac{t}{\sqrt{2}}\right) + C = \frac{1}{\sqrt{2}} \arctan\left(\frac{\tan x}{\sqrt{2}}\right) + C,$$

where $C \in \mathbb{R}$.

# Chapter 15

# The Solutions to the 2023 MIT Integration Bee, Qualifying Test

$$I_1 = \int x^{1/\ln x} dx$$

**Solution.**

Remember that $a^b = e^{b\ln a}$, $\forall a > 0, b \in \mathbb{R}$. Thus, for the given integral we can write the following

$$I_1 = \int e^{(1/\ln x)\ln x} dx = \int e dx = ex + C,$$

where $C \in \mathbb{R}$.

$$I_2 = \int \operatorname{sech} x dx$$

**Solution.**

For the given integral we can write the following

$$I_2 = \int \frac{dx}{\cosh x} = 2 \int \frac{dx}{e^x + e^{-x}} = 2 \int \frac{dx}{e^{-x}(1 + (e^x)^2)}$$
$$= 2 \int \frac{e^x}{1 + (e^x)^2} dx = 2\arctan(e^x) + C,$$

where $C \in \mathbb{R}$.

$$I_3 = \int \frac{e^x}{(1 + e^x) \ln (1 + e^x)} \, dx$$

**Solution.**

Let us use the substitution $e^x = t$, so that $e^x dx = dt$. This gives the following

$$I_3 = \int \frac{dt}{(1 + t) \ln(1 + t)} = \int \frac{(\ln(1 + t))'}{\ln(1 + t)} \, dt = \ln (\ln(1 + t)) + C$$
$$= \ln (\ln(1 + e^x)) + C,$$

where $C \in \mathbb{R}$.

$$I_4 = \int \left(1 + x + x^2 + x^3 + x^4\right) \left(1 - x + x^2 - x^3 + x^4\right) dx$$

**Solution.**

For the given integral we can write the following

$$I_4 = \int \left((1 + x) + x^2(1 + x) + x^4\right) \left((1 - x) + x^2(1 - x) + x^4\right) dx$$
$$= \int \left((1 + x)(1 + x^2) + x^4\right) \left((1 - x)(1 + x^2) + x^4\right) dx$$
$$= \int \left((1 - x^2)(1 + x^2)(1 + x^2) + x^4(1 + x)(1 + x^2)\right.$$
$$\left. + x^4(1 - x)(1 + x^2) + x^8\right) dx$$
$$= \int \left((1 - x^4)(1 + x^2) + x^4(1 + x^2)(1 + x + 1 - x) + x^8\right) dx$$
$$= \int \left(1 + x^2 + x^4 + x^6 + x^8\right) dx = x + \frac{x^3}{3} + \frac{x^5}{5} + \frac{x^7}{7} + \frac{x^9}{9} + C,$$

where $C \in \mathbb{R}$.

$$I_5 = \int_0^4 \left(\frac{x}{5}\right) dx$$

**Solution 1.**

For any $x \in [0, 4]$, we have $\left(\frac{x}{5}\right) = 0$. Thus $I_5 = \int_0^4 0 \, dx = 0$.

**Solution 2.**

For the given integral we can write the following

$$I_5 = \int_0^4 \frac{x!}{5!(x-5)!}dx = \int_0^4 \frac{x(x-1)(x-2)(x-3)(x-4)(x-5)!}{5!(x-5)!}dx$$

$$= \frac{1}{5!}\int_0^4 x(x-1)(x-2)(x-3)(x-4)dx$$

$$= \frac{1}{5!}\int_0^4 \left(x^5 - 10x^4 + 35x^3 - 50x^2 + 24x\right)dx$$

$$= \frac{1}{5!}\left[\frac{x^6}{6} - 2x^5 + \frac{35}{4}x^4 - \frac{50}{3}x^3 + 12x^2\right]_0^4$$

$$= \frac{1}{5!}\left(\frac{4^6}{6} - 2(4^5) + \frac{35}{4}4^4 - \frac{50}{3}4^3 + 12(4^2)\right) = 0.$$

$$\boxed{I_6 = \int (x + \sin x + x\cos x + \sin x \cos x)dx}$$

**Solution 1.**

For the given integral we can write the following

$$I_6 = \int \left(x(1 + \cos x) + \sin x(1 + \cos x)\right)dx = \int (1 + \cos x)(x + \sin x)dx$$

$$= \int (x + \sin x)(x + \sin x)'dx = \frac{1}{2}(x + \sin x)^2 + C,$$

where $C \in \mathbb{R}$.

**Solution 2.**

For the given integral we can write the following

$$I_6 = \int xdx + \int \sin xdx + \int \sin x \cos xdx + \underbrace{\int x \cos xdx}_{:=J}.$$

For the integral $J$, by using the integration by parts (assume $u = x, dv = \cos xdx$), we find

$$J = x\sin x - \int \sin xdx = x\sin x + \cos x + C,$$

where $C \in \mathbb{R}$. Thus, we have

$$I_6 = \frac{x^2}{2} - \cos x + \frac{1}{2} \sin^2 x + x \sin x + \cos x + C$$

$$= \frac{x^2}{2} + \frac{1}{2} \sin^2 x + x \sin x + C$$

$$= \frac{1}{2} \left( x^2 + 2x \sin x + \sin^2 x \right) + C = \frac{1}{2} \left( x + \sin x \right)^2 + C.$$

$$\boxed{I_7 = \int \left( \sin^2 x + \cos^2 x + \tan^2 x + \cot^2 x + \sec^2 x + \csc^2 x \right) dx}$$

**Solution.**

For the given integral we can write the following

$$I_7 = \int \left( 1 + \frac{\sin^2 x}{\cos^2 x} + \frac{\cos^2 x}{\sin^2 x} + \frac{1}{\cos^2 x} + \frac{1}{\sin^2 x} \right) dx$$

$$= \int \left( 1 + \frac{1 - \cos^2 x}{\cos^2 x} + \frac{1 - \sin^2 x}{\sin^2 x} + \frac{1}{\cos^2 x} + \frac{1}{\sin^2 x} \right) dx$$

$$= \int \left( 1 + \frac{1}{\cos^2 x} - 1 + \frac{1}{\sin^2 x} - 1 + \frac{1}{\cos^2 x} + \frac{1}{\sin^2 x} \right) dx$$

$$= \int \left( -1 + \frac{2}{\cos^2 x} + \frac{2}{\sin^2 x} \right) dx = -x + 2 \tan x - 2 \cot x + C,$$

where $C \in \mathbb{R}$.

$$\boxed{I_8 = \int_0^{2\pi} \lfloor 2023 \sin x \rfloor dx}$$

**Solution.**

Let us calculate the general form of the integral $I_8$, which has the following form

$$J_n = \int_0^{2\pi} \lfloor n \sin x \rfloor dx, \quad \forall n \in \mathbb{N}. \tag{15.1}$$

For $n = 1$, we have $J_1 = \int_0^{2\pi} \lfloor \sin x \rfloor dx$, and in the interval $[0, 2\pi]$, we have

$$0 \leqslant \sin x < 1, \ \forall x \in [0, \pi] \setminus \pi/2 \implies \lfloor \sin x \rfloor = 0, \ \forall x \in [0, \pi] \setminus \{\pi/2\},$$
$$-1 \leqslant \sin x < 0, \ \forall x \in ]\pi, 2\pi[ \implies \lfloor \sin x \rfloor = -1, \ \forall x \in ]\pi, 2\pi[.$$

Thus we have (see Fig. 15.1)

$$J_1 = \int_\pi^{2\pi} (-1) dx = -(2\pi - \pi) = -\pi.$$

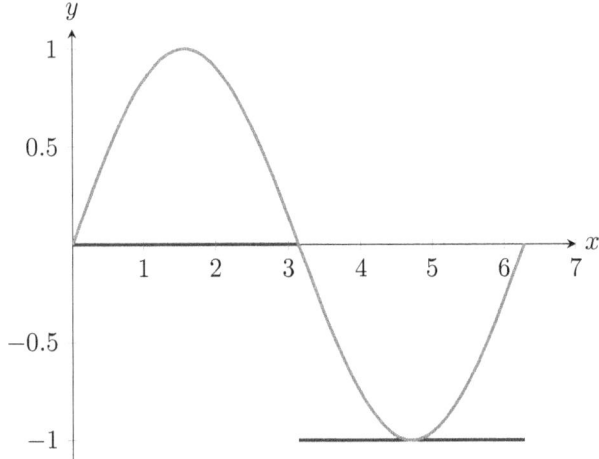

Figure 15.1: The graph of $\sin x$ and $\lfloor \sin x \rfloor$ in $[0, 2\pi]$.

For $n = 2$, we have $J_2 = \int_0^{2\pi} \lfloor 2 \sin x \rfloor dx$. Let us find $x \in [0, 2\pi]$, such that

$$2 \sin x = m; \quad m = -2, -1, 0, 1, 2.$$
$$\implies x = \arcsin(m/2); \quad m = -2, -1, 0, 1, 2.$$

When

$$
\begin{aligned}
m = 0 \quad &\implies \quad x = 0, \pi, 2\pi. \\
m = 1 \quad &\implies \quad x = \pi/6, 5\pi/6. \\
m = 2 \quad &\implies \quad x = \pi/2. \\
m = -1 \quad &\implies \quad x = 7\pi/6, 11\pi/6. \\
m = -2 \quad &\implies \quad x = 3\pi/2.
\end{aligned}
$$

For these founded values in $[0, 2\pi]$, such that $\sin x \in \{-2, -1, 0, 1, 2\}$, we get

**(1)**

$$0 \leqslant \sin x < 1/2, \ \forall x \in [0, \pi/6[ \implies 0 \leqslant 2 \sin x < 1, \ \forall x \in [0, \pi/6[,$$
$$\implies \lfloor 2 \sin x \rfloor = 0, \ \forall x \in [0, \pi/6[.$$

**(2)**

$$1/2 \leqslant \sin x < 1, \ \forall x \in [\pi/6, 5\pi/6] \setminus \{\pi/2\},$$
$$\implies 1 \leqslant 2 \sin x < 2, \ \forall x \in [\pi/6, 5\pi/6] \setminus \{\pi/2\},$$
$$\implies \lfloor 2 \sin x \rfloor = 1, \ \forall x \in [\pi/6, 5\pi/6] \setminus \{\pi/2\}.$$

(3)

$$0 \leqslant \sin x < 1/2, \ \forall x \in \ ]5\pi/6, \pi] \implies 0 \leqslant 2 \sin x < 1, \ \forall x \in \ ]5\pi/6, \pi],$$
$$\implies \lfloor 2 \sin x \rfloor = 0, \ \forall x \in \ ]5\pi/6, \pi].$$

(4)

$$-1/2 \leqslant \sin x < 0, \ \forall x \in \ ]\pi, 7\pi/6[ \implies -1 \leqslant 2 \sin x < 0, \ \forall x \in \ ]\pi, 7\pi/6[,$$
$$\implies \lfloor 2 \sin x \rfloor = -1, \ \forall x \in \ ]\pi, 7\pi/6[.$$

(5)

$$-1 \leqslant \sin x < -1/2, \ \forall x \in \ ]7\pi/6, 11\pi/6[,$$
$$\implies -2 \leqslant 2 \sin x < -1, \ \forall x \in \ ]7\pi/6, 11\pi/6[,$$
$$\implies \lfloor 2 \sin x \rfloor = -2, \ \forall x \in \ ]7\pi/6, 11\pi/6[.$$

(6)

$$-1/2 \leqslant \sin x < 0, \ \forall x \in \ [11\pi/6, 2\pi[,$$
$$\implies -1 \leqslant 2 \sin x < 0, \ \forall x \in \ [11\pi/6, 2\pi[,$$
$$\implies \lfloor 2 \sin x \rfloor = -1, \ \forall x \in \ [11\pi/6, 2\pi[.$$

See Fig. 15.2. Thus, we have

$$J_2 = \int_{\pi/6}^{5\pi/6} dx - \int_{\pi}^{7\pi/6} dx - 2 \int_{7\pi/6}^{11\pi/6} dx - \int_{11\pi/6}^{2\pi} dx$$
$$= \left( \frac{5\pi}{6} - \frac{\pi}{6} \right) - \left( \frac{7\pi}{6} - \pi \right) - 2 \left( \frac{11\pi}{6} - \frac{7\pi}{6} \right) - \left( 2\pi - \frac{11\pi}{6} \right)$$
$$= -\pi.$$

Now for the general case $J_n$ in (15.1). Let us take $\alpha, \beta \in ]0, \pi[$. For any $x \in [\alpha, \beta[$, there exist $t \in \{0, 1, 2, \ldots, n-1\}$, such that $\lfloor n \sin x \rfloor = t \ \forall x \in [\alpha, \beta[$ (See Fig. 15.3). Because the graph of the function $f(x) := n \sin x$, i.e. $\mathcal{C}_f = \{(x, y) \in \mathbb{R}^n : y = f(x), \forall x \in [0, 2\pi]\}$, has a point $(\pi, 0)$ as a center of the symmetry on $[0, 2\pi]$ for different $n \in \mathbb{N}$ (see Fig. 15.1 for $n = 1$, and Fig. 15.2 for $n = 2$), we can find an interval $[\alpha + \pi, \beta + \pi[ \subset [\pi, 2\pi[$, such that

$$\lfloor n \sin x \rfloor = -(t+1), \quad \forall x \in [\alpha + \pi, \beta + \pi[. \tag{15.2}$$

Thus, we have

$$\int_{\alpha}^{\beta} \lfloor n \sin x \rfloor dx + \int_{\alpha+\pi}^{\beta+\pi} \lfloor n \sin x \rfloor dx = \int_{\alpha}^{\beta} t \, dx - \int_{\alpha+\pi}^{\beta+\pi} (t+1) dx$$
$$= t(\beta - \alpha) - (t+1)(\beta + \pi - \alpha - \pi)$$
$$= \alpha - \beta. \tag{15.3}$$

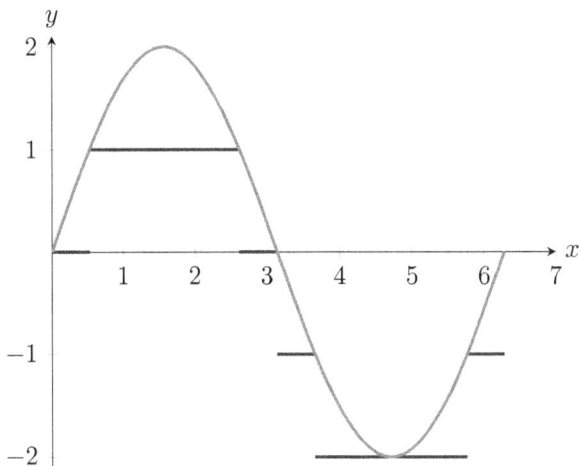

Figure 15.2: The graph of $2\sin x$ and $\lfloor 2\sin x\rfloor$ in $[0, 2\pi]$.

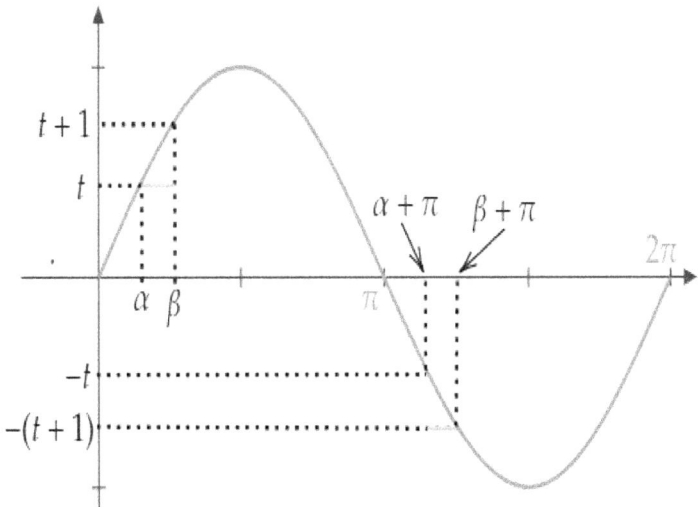

Figure 15.3: Illustration for (15.2), (15.3) and the connected formulas.

Now, let us use the previous result to calculate the integral (15.1).

Let $a_1, \ldots, a_N$ the solutions of the equation

$$\sin x = \frac{m}{n}, \quad \text{where } m = 0, 1, 2, \ldots, n$$

in the interval $[0, \pi]$, such that $\{a_1, \ldots, a_N\}$ represents a partition of $[0, \pi]$, i.e. $a_1 = 0 < a_2 < \ldots < a_N = \pi$, and

$$\bigcup_{i=1}^{N-1} [a_i, a_{i+1}] = [0, \pi], \quad ]a_j, a_{j+1}[ \cap ]a_{j'}, a_{j'+1}[ = \emptyset,$$

$\forall j, j' \in \{1, \ldots, N-1\}$ and $i \neq j$. For any $x \in [a_i, a_{i+1}[$ ($\forall i = 1, 2, \ldots, N-1$), there exist $t_i \in \{0, 1, 2, \ldots, n-1\}$, such that

$$\lfloor n \sin x \rfloor = t_i, \quad \forall x \in [a_i, a_{i+1}[.$$

Now, for the integral (15.1), we get

$$J_n = \int_0^{2\pi} \lfloor n \sin x \rfloor dx = \sum_{i=1}^{N-1} \left( \int_{a_i}^{a_{i+1}} \lfloor n \sin x \rfloor dx + \int_{a_i+\pi}^{a_{i+1}+\pi} \lfloor n \sin x \rfloor dx \right)$$

$$= \sum_{i=1}^{N-1} \left( \int_{a_i}^{a_{i+1}} t_i dx + \int_{a_i+\pi}^{a_{i+1}+\pi} -(t_i+1) dx \right)$$

$$= \sum_{i=1}^{N-1} \left( (a_{i+1} - a_i)t_i - (t_i+1)(a_{i+1} + \pi - a_i - \pi) \right) = \sum_{i=1}^{N-1} (a_i - a_{i+1})$$

$$= a_1 - a_2 + a_2 - a_3 + a_3 - \ldots + a_{N-1} - a_N = a_1 - a_N = 0 - \pi = -\pi.$$

Therefore, for the given integral we have

$$I_8 = \int_0^{2\pi} \lfloor 2023 \sin x \rfloor dx = -\pi.$$

$$\boxed{I_9 = \int (1 + 2 \ln x) e^{(\ln x)^2} dx}$$

**Solution.**

For the given integral we can write the following

$$I_9 = 2 \int \ln x e^{(\ln x)^2} dx + \underbrace{\int e^{(\ln x)^2} dx}_{:=J}.$$

For the integral $J$, by using the integration by parts, let us assume

$$u = e^{(\ln x)^2} \implies du = \frac{2\ln x}{x}e^{(\ln x)^2}dx, \quad dv = dx \implies v = x.$$

Thus, we have

$$I_9 = 2\int \ln x e^{(\ln x)^2}dx + xe^{(\ln x)^2} - 2\int x\frac{\ln x}{x}e^{(\ln x)^2}dx$$

$$= 2\int \ln x e^{(\ln x)^2}dx + xe^{(\ln x)^2} - 2\int \ln x e^{(\ln x)^2}dx = xe^{(\ln x)^2} + C,$$

where $C \in \mathbb{R}$.

$$I_{10} = \int (1-x)^3 + \left(x - x^2\right)^3 + \left(x^2 - 1\right)^3 - 3(1-x)\left(x - x^2\right)\left(x^2 - 1\right)dx$$

**Solution.**

For the given integral we can write the following

$$I_{10} = \int \Big((1-x)^3 + x^3(1-x)^3 - (1-x)^3(1+x)^3$$

$$+ 3(1-x)x(1-x)(1-x^2)\Big)dx$$

$$= \int \left((1-x)^3 + x^3(1-x)^3 - (1-x)^3(1+x)^3 + 3x(1-x)^3(1+x)\right)dx$$

$$= \int (1-x)^3\left(1 + x^3 - (1+x)^3 + 3x(1+x)\right)dx$$

$$= \int (1-x)^3\left((1+x)(1 - x + x^2) - (1+x)^3 + 3x(1+x)\right)dx$$

$$= \int (1-x)^3\left((1+x)(1 - x + x^2 - (1+x)^2 + 3x)\right)dx$$

$$= \int (1-x)^3\left((1+x)(1 - x + x^2 - 1 - 2x - x^2 + 3x)\right)dx$$

$$= \int 0 dx = C,$$

where $C \in \mathbb{R}$.

$$I_{11} = \int_{-2023}^{2023} \underbrace{||||||x| - 1| - 1|\ldots| - 1|}_{2023(-1)'s} dx$$

**Solution.**

Let us calculate the general form of $I_{11}$, which has the following from

$$J_n = \int_{-n}^{n} \underbrace{|||||x| - 1| - 1|\ldots| - 1|}_{n(-1)'s}\, dx, \quad n \in \mathbb{N} = \{1, 2, \ldots\}.$$

The integrand $f_n(x) := \underbrace{|||||x| - 1| - 1|\ldots| - 1|}_{n(-1)'s}$, $\forall x \in [-n, n]$, is an even function. Thus, we can write

$$J_n = 2\int_{0}^{n} \underbrace{|||||x| - 1| - 1|\ldots| - 1|}_{n(-1)'s}\, dx, \quad n \in \mathbb{N} = \{1, 2, \ldots\}. \qquad (15.4)$$

For $n = 1$, we have

$$J_1 = 2\int_0^1 ||x| - 1|\, dx = 2\int_0^1 |x - 1|dx = 2\int_0^1 (1 - x)dx = 2\left[x - \frac{x^2}{2}\right]_0^1$$
$$= 2\left(1 - \frac{1}{2}\right) = 1.$$

For $n = 2$, we have

$$J_2 = 2\int_0^2 |||x| - 1| - 1|\, dx = 2\int_0^2 ||x - 1| - 1|\, dx$$
$$= 2\left(\int_0^1 ||x - 1| - 1|dx + \int_1^2 ||x - 1| - 1|dx\right)$$
$$= 2\left(\int_0^1 |1 - x - 1|dx + \int_1^2 |x - 1 - 1|dx\right)$$
$$= 2\left(\int_0^1 xdx + \int_1^2 (2 - x)dx\right) = 2\left(\left[\frac{x^2}{2}\right]_0^1 + \left[2x - \frac{x^2}{2}\right]_1^2\right) = 2.$$

For $n = 3$, we have

$$J_3 = 2\int_0^3 |||||x| - 1| - 1| - 1|\, dx.$$

From a geometrical viewpoint, the integral $\int_0^3 |||||x| - 1| - 1| - 1|\, dx$ represents the sum of three right triangles. The base and height of each triangle are equal to 1, see Fig. 15.4 and Fig. 15.5. Thus,

$$J_3 = (2)(3)\left(\frac{1}{2}(1)(1)\right) = 3.$$

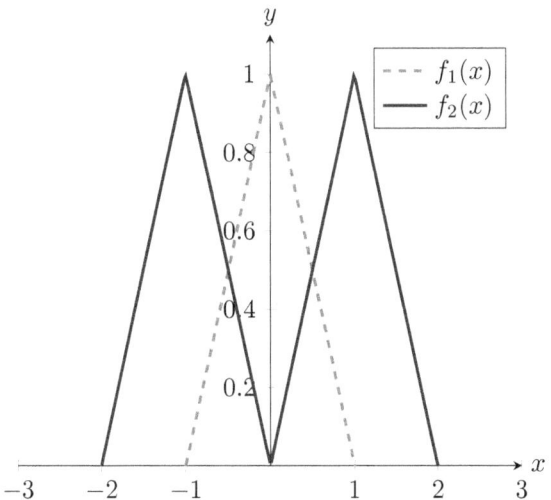

Figure 15.4: The graphs of $f_1(x) := ||x| - 1|$ and $f_2(x) := |||x| - 1| - 1|$.

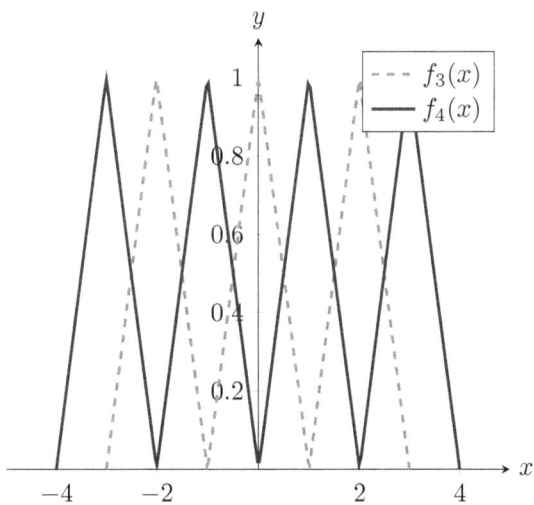

Figure 15.5: The graphs of $f_3(x) := ||||x| - 1| - 1| - 1|$ and $f_4(x) := |||||x| - 1| - 1| - 1| - 1|$.

By similar arguments and by induction, we can find that the integral on the right side of (15.4), represents the sum of $n$ right triangles. The base and height of each triangle are equal to 1. Thus,

$$J_n = 2n \left( \frac{1}{2}(1)(1) \right) = n.$$

Therefore, for the given integral $I_{11}$, we find

$$I_{11} = \int_{-2023}^{2023} \underbrace{|||||x| - 1| - 1| \ldots | - 1|}_{2023(-1)'s} dx = J_{2023} = 2023.$$

$$\boxed{I_{12} = \int \left( \sin^6 x + \cos^6 x + 3 \sin^2 x \cos^2 x \right) dx}$$

**Solution.**

For the given integral, we have

$$\sin^2 x + \cos^2 x = 1,$$
$$\implies \left( \sin^2 x + \cos^2 x \right)^3 = 1,$$
$$\implies \sin^6 x + 3 \sin^4 x \cos^2 x + 3 \sin^2 x \cos^4 x + \cos^6 x = 1,$$
$$\implies \sin^6 x + \cos^6 x + 3 \sin^2 x \cos^2 x (\sin^2 x + \cos^2 x) = 1,$$
$$\implies \sin^6 x + \cos^6 x + 3 \sin^2 x \cos^2 x = 1.$$

Therefore

$$I_{12} = \int dx = x + C,$$

where $C \in \mathbb{R}$.

$$\boxed{I_{13} = \int (x + e + 1)x^e e^x dx}$$

**Solution 1.**

We have

$$(x + e + 1)x^e e^x = x^{e+1} e^x + (e + 1)x^e e^x,$$

and

$$\left( x^{e+1} e^x \right)' = x^{e+1} e^x + (e + 1)x^e e^x.$$

Thus, we have

$$I_{13} = \int x^{e+1} e^x + (e + 1)x^e e^x dx = \int \left( x^{e+1} e^x \right)' dx = x^{e+1} e^x + C,$$

where $C \in \mathbb{R}$.

**Solution 2.**

For the given integral we can write the following

$$I_{13} = \underbrace{\int xx^e e^x dx}_{:=J} + (e+1) \int x^e e^x dx.$$

For the integral $J$, by using the integration by parts, let us assume

$$u = x^{e+1} \implies du = (e+1)x^e dx, \quad dv = e^x dx \implies v = e^x.$$

Thus, we have

$$I_{13} = x^{e+1}e^x - (e+1)\int x^e e^x dx + (e+1)\int x^e e^x dx = x^{e+1}e^x + C,$$

where $C \in \mathbb{R}$.

$$I_{14} = \int_0^1 \left( \frac{x^2}{2-x^2} + \sqrt{\frac{2x}{x+1}} \right) dx$$

**Solution.**

For the given integral, let

$$J_1 := \int_0^1 \frac{x^2}{2-x^2} dx, \quad J_2 := \int_0^1 \sqrt{\frac{2x}{x+1}} dx \implies I_{14} = J_1 + J_2.$$

For the integral $J_1$, we have

$$J_1 = \int_0^1 \left( -1 + \frac{2}{2-x^2} \right) dx = -1 + \int_0^1 \frac{2}{2-x^2} dx.$$

For the integral $J_2$, let us use the substitution $\frac{2x}{x+1} = t^2$, so that $x = \frac{t^2}{2-t^2}$, and $dx = \frac{4t}{(2-t^2)^2} dt$. If $x = 0$, then $t = 0$, and if $x = 1$, then $t = 1$. This gives the following

$$J_2 = 4 \int_0^1 t \frac{t}{(2-t^2)^2} dt.$$

Now, by using the integration by parts, let us assume

$$u = t \implies du = dt, \quad dv = \frac{t}{(2-t^2)^2} dt \implies v = \frac{1}{2(2-t^2)}.$$

Thus, we have

$$J_2 = 4 \left( \left[ \frac{t}{2(2 - t^2)} \right]_0^1 - \frac{1}{2} \int_0^1 \frac{dt}{2 - t^2} \right) = 2 - 2 \int_0^1 \frac{dt}{2 - t^2}.$$

Therefore, for the given integral $I_{14}$, we have

$$I_{14} = J_1 + J_2 = -1 + 2 \int_0^1 \frac{dx}{2 - x^2} + 2 - 2 \int_0^1 \frac{dx}{2 - x^2} = 1.$$

$$\boxed{I_{15} = \int \frac{1 + 2x^{2022}}{x + x^{2023}} \, dx}$$

**Solution.**

For the given integral we can write the following

$$
\begin{aligned}
I_{15} &= \int \frac{1 + x^{2022}}{x(1 + x^{2022})} \, dx + \int \frac{x^{2022}}{x(1 + x^{2022})} \, dx \\
&= \int \frac{dx}{x} + \int \frac{x^{2021}}{1 + x^{2022}} \, dx \\
&= \int \frac{dx}{x} + \frac{1}{2022} \int \frac{2022 x^{2021}}{1 + x^{2022}} \, dx \\
&= \ln|x| + \frac{1}{2022} \ln\left(1 + x^{2022}\right) + C \\
&= \frac{1}{2022} \left(2022 \ln|x| + \ln\left(1 + x^{2022}\right)\right) + C \\
&= \frac{1}{2022} \ln\left(x^{2022}\left(1 + x^{2022}\right)\right) + C \\
&= \frac{1}{2022} \ln\left(x^{2022} + x^{4044}\right) + C,
\end{aligned}
$$

where $C \in \mathbb{R}$.

$$\boxed{I_{16} = \int \left(3 \sin(20x) \cos(23x) + 20 \sin(43x)\right) dx}$$

**Solution.**

For the given integral we can write the following

$$I_{16} = \int \left( \frac{3}{2} \left( \sin(43x) + \sin(-3x) \right) + 20 \sin(43x) \right) dx$$

$$= \int \left( \frac{3}{2} \sin(43x) - \frac{3}{2} \sin(3x) + 20 \sin(43x) \right) dx$$

$$= \int \left( \frac{43}{2} \sin(43x) - \frac{3}{2} \sin(3x) \right) dx$$

$$= \frac{1}{2} \left( \cos(3x) - \cos(43x) \right) + C$$

$$= \frac{1}{2} \left( -2 \sin \left( \frac{3x + 43x}{2} \right) \sin \left( \frac{3x - 43x}{2} \right) \right) + C$$

$$= \sin(23x) \sin(20x) + C,$$

where $C \in \mathbb{R}$.

$$\boxed{I_{17} = \int_0^1 \prod_{k=0}^{\infty} \left( \frac{1}{1 + x^{2^k}} \right) dx}$$

**Solution.**

We have

$$\prod_{k=0}^{\infty} \left( \frac{1}{1 + x^{2^k}} \right) = \frac{1}{\prod_{k=0}^{\infty} \left( 1 + x^{2^k} \right)}.$$

Now, let us find a formula for $\prod_{k=0}^{\infty} \left( 1 + x^{2^k} \right)$ as a function of $x$, where $0 < x < 1$.

For this product we have

$$P_n := \prod_{k=0}^{n} \left( 1 + x^{2^k} \right) = (1 + x) \left( 1 + x^2 \right) \left( 1 + x^4 \right) \left( 1 + x^{2^{k-1}} \right) \left( 1 + x^{2^k} \right).$$

By multiplying both sides of the previous equality by $1 - x \neq 0$, we find

$$(1 - x) P_n = (1 - x)(1 + x) \left( 1 + x^2 \right) \left( 1 + x^4 \right) \cdots \left( 1 + x^{2^{n-1}} \right) \left( 1 + x^{2^n} \right)$$

$$= \left( 1 - x^2 \right) \left( 1 + x^2 \right) \left( 1 + x^4 \right) \cdots \left( 1 + x^{2^{n-1}} \right) \left( 1 + x^{2^n} \right)$$

$$= \left( 1 - x^4 \right) \left( 1 + x^4 \right) \left( 1 + x^8 \right) \cdots \left( 1 + x^{2^{n-1}} \right) \left( 1 + x^{(2^n)} \right)$$

$$= \ldots = \left( 1 - x^{2^n} \right) \left( 1 + x^{2^n} \right) = \left( 1 + x^{2^{n+1}} \right).$$

Thus, $P_n = \frac{1+x^{2^{n+1}}}{1-x}$. Because of $0 < x < 1$, we have $\lim_{n\to\infty} x^{2^{n+1}} = 0$, and therefore the product $\prod_{k=0}^{\infty} \left(1 + x^{2^k}\right)$ is convergence, where

$$\prod_{k=0}^{\infty}\left(1+x^{2^k}\right) = \lim_{n\to\infty} \prod_{k=0}^{n}\left(1+x^{2^k}\right) = \lim_{n\to\infty} P_n = \lim_{n\to\infty} \frac{1+x^{2^{n+1}}}{1-x} = \frac{1}{1-x}.$$

Therefore, for the given integral $I_{17}$, we get

$$I_{17} = \int_0^1 (1-x)dx = \left[x - \frac{x^2}{2}\right]_0^1 = \left(1 - \frac{1}{2}\right) = \frac{1}{2}.$$

$$I_{18} = \int \frac{\sin x}{2e^x + \cos x + \sin x}dx$$

**Solution.**

For the given integral we can write the following

$$I_{18} = \frac{1}{2}\int \frac{2\sin x + \cos x - \cos x + 2e^x - 2e^x}{2e^x + \cos x + \sin x}dx$$

$$= \frac{1}{2}\left(\int \frac{\sin x + \cos x + 2e^x}{2e^x + \cos x + \sin x}dx + \int \frac{\sin x - \cos x - 2e^x}{2e^x + \cos x + \sin x}dx\right)$$

$$= \frac{1}{2}\left(\int dx - \int \frac{\cos x - \sin x + 2e^x}{2e^x + \cos x + \sin x}dx\right)$$

$$= \frac{1}{2}\left(x - \ln\left|2e^x + \cos x + \sin x\right|\right) + C,$$

where $C \in \mathbb{R}$.

$$I_{19} = \int \frac{\ln(x/\pi)}{(\ln x)^{\ln(e\pi)}}dx$$

**Solution.**

For the given integral we can write the following

$$I_{19} = \int \frac{\ln x - \ln \pi}{(\ln x)^{1+\ln \pi}}dx = \int \frac{\ln x}{\ln x(\ln x)^{\ln \pi}}dx - \int \frac{\ln \pi}{(\ln x)^{1+\ln \pi}}dx$$

$$= \int \frac{dx}{(\ln x)^{\ln \pi}} - \ln \pi \int \frac{dx}{(\ln x)^{1+\ln \pi}}$$

$$= \underbrace{\int (\ln x)^{-\ln \pi}dx}_{:=J} - \ln \pi \int \frac{dx}{(\ln x)^{1+\ln \pi}}.$$

For the integral $J$, by using the integration by parts, let us assume

$$u = (\ln x)^{-\ln \pi} \implies du = -\ln \pi (\ln x)^{-1-\ln \pi} \frac{dx}{x}, \quad dv = dx \implies v = x.$$

Thus, we have

$$I_{19} = x(\ln x)^{-\ln \pi} + \ln \pi \int x(\ln x)^{-1-\ln \pi} \frac{dx}{x} - \ln \pi \int (\ln x)^{-1-\ln \pi} dx$$

$$= x(\ln x)^{-\ln \pi} + \ln \pi \int (\ln x)^{-1-\ln \pi} dx - \ln \pi \int (\ln x)^{-1-\ln \pi} dx$$

$$= x(\ln x)^{-\ln \pi} + C = \frac{x}{(\ln x)^{\ln \pi}} + C,$$

where $C \in \mathbb{R}$.

$$I_{20} = \int_{-3/2}^{-1/2} \left( x^5 + 5x^4 + 10x^3 + 8x^2 + x \right) dx$$

**Solution.**
We can calculate this integral simply as follows

$$I_{20} = \left[ \frac{x^6}{6} + x^5 + \frac{5}{2}x^4 + \frac{8}{3}x^3 + \frac{x^2}{2} \right]_{-3/2}^{-1/2}$$

$$= \left( \frac{1}{6(2^6)} + \frac{1}{2^5} + \frac{5}{2} \cdot \frac{1}{2^4} + \frac{8}{3} \cdot \frac{1}{2^3} + \frac{1}{2(2^2)} \right)$$

$$- \left( \frac{3^6}{6(2^6)} - \frac{3^5}{2^5} + \frac{5}{2} \cdot \frac{3^4}{2^4} - \frac{8}{3} \cdot \frac{3^3}{2^3} + \frac{3^2}{2(2^2)} \right)$$

$$= \frac{5}{6}.$$

# References

1. Thomas George B., Maurice D. Weir, and Joel Hass: Thomas' Calculus. Twelfth Edition, Addison-Wesley. 2010.

2. Anton H., Bivens I., Davis S., and Polaski T.: Calculus early transcendentals. Ninth edition, JOHN WILEY & SONS, INC. 2009.

3. Michael Spivak: Calculus. Third Edition, Publish or Perish, Inc. 1994.

4. Royden H.L.: *Real Analysis*. Prentice Hall, 1988.

5. Wikipedia, the free encyclopedia. Dominated convergence theorem, https://en.wikipedia.org/wiki/Dominated_convergence_theorem

6. Wikipedia, the free encyclopedia. Inverse trigonometric functions, https://en.wikipedia.org/wiki/Inverse_trigonometric_functions

7. 2010 Integration Bee, Qualifying Test. January 25, 2010. https://math.mit.edu/~yyao1/pdf/qualifying_round_2010_test.pdf

8. 2011 Integration Bee, Qualifying Test. January 14, 2011. https://math.mit.edu/~yyao1/pdf/qualifying_round_2011_test.pdf

9. 2012 Integration Bee, Qualifying Test. January 13, 2012. https://math.mit.edu/~yyao1/pdf/qualifying_round_2012_test.pdf

10. 2013 Integration Bee, Qualifying Test. 11 January 2013. https://math.mit.edu/~yyao1/pdf/qualifying_round_2013_answers.pdf

11. 2014 Integration Bee, Qualifying Test. 21 January 2014. https://math.mit.edu/~yyao1/pdf/qualifying_round_2014_test.pdf

12. 2015 Integration Bee, Qualifying Test. 20 January 2015. https://math.mit.edu/~yyao1/pdf/qualifying_round_2015_test.pdf

13. 2016 Integration Bee, Qualifying Test. 19 January 2016. https://math.mit.edu/~yyao1/pdf/qualifying_round_2016_test.pdf

14. 2017 Integration Bee, Qualifying Test. 24 January 2017. https://math.mit.edu/~yyao1/pdf/qualifying_round_2017_test.pdf

15. 2018 Integration Bee, Qualifying Test. 23 January 2018. https://math.mit.edu/~yyao1/pdf/qualifying_round_2018_test.pdf

16. 2019 Integration Bee, Qualifying Test. 29 January 2019. https://math.mit.edu/~yyao1/pdf/qualifying_round_2019_test.pdf

17. 2021 Integration Bee, Qualifying Test. 21 January 2020. https://math.mit.edu/~yyao1/pdf/qualifying_round_2020_test.pdf

18. 2022 Integration Bee, Qualifying Test. 18 January 2022. https://math.mit.edu/~yyao1/pdf/qualifying_round_2022_test.pdf

19. 2023 Integration Bee, Qualifying Test. 24 January 2023. https://math.mit.edu/~yyao1/pdf/qualifying_round_2023_test.pdf

www.ingramcontent.com/pod-product-compliance
Lightning Source LLC
Chambersburg PA
CBHW060827220526
45466CB00003B/1003